Fruitful Legacy:

A Historic Context of Orchards in the United States, with Technical Information for Registering Orchards in the National Register of Historic Places

Susan A. Dolan

2009

National Park Service
Olmsted Center for Landscape Preservation
Pacific West Regional Office – Cultural Resources
Park Historic Structures and Cultural Landscapes Program

Library of Congress Cataloging-in-Publication Data

Dolan, Susan A. 1967-

Fruitful legacy : a historic context of orchards in the United States, with technical information for registering orchards in the National Register of Historic Places / Susan A. Dolan.

p. cm.

Includes bibliographical references.

ISBN 0-16-082127-4 (pbk.)

1. Fruit-culture--United States--History. 2. Orchards--United States--History. 3. Historic sites--United States. I. United States. National Park Service. Park Historic Structures and Cultural Landscapes Program. II. Title.

SB355.D65 2009

634.0973--dc22

200902351

ISBN 978-0-16-082127-1

90000

9 780160 821271

For sale by the Superintendent of Documents, U.S. Government Printing Office
Internet: bookstore.gpo.gov Phone: toll free (866) 512-1800; DC area (202) 512-1800
Fax: (202) 512-2104 Mail: Stop IDCC, Washington, DC 20402-0001

ISBN 978-0-16-082127-1

"It is remarkable how closely the history of the apple tree is connected with that of man."

Henry David Thoreau / Essay *"Wild Apples"* / 1862

"... I've been working on an idea that still has a way to go — to use the apples grown in some National Park Service historic orchards as a means of informing visitors about these special fruits and the importance of protecting the genetic variety they represent. One of my thoughts has involved a way to sell visitors the individual fruits wrapped in tissue paper, each tissue having the history of the orchard printed on it. I see this as a way to condense into a simple form the profoundly important topic of biological diversity."

NPS Director William Penn Mott, Jr.
"The Director's Report" / Courier / June 1988

CONTENTS

Chapter 4

Part II: Technical Information for Registering Orchards in the National Register of Historic Places

Chapter 5

Part III: Further Information

List of Figures

Introduction Figures

Chapter 1 Figures

Chapter 2 Figures

Chapter 3 Figures

Chapter 4 Figures

Chapter 5 Figures

List of Tables

Chapter 3 Tables

Chapter 4 Tables

Chapter 5 Tables

Acknowledgments

This project began as a thoughtful discussion en route to one of the most impressive historic orchards in the national park system—the Buckner Orchard in North Cascades National Park Service Complex in Washington State. The year was 2000 and my supervisor, Cathy Gilbert (then Program Lead for the Cultural Landscapes Program in the Pacific West Region) and myself (a Historical Landscape Architect and relative newcomer to the agency) were paying a visit to the historic Buckner Orchard with two colleagues from the East Coast: Bob Page (former Park Cultural Landscapes Program Manager in the Washington, DC office and now Director of the Olmsted Center for Landscape Preservation and Chief of Cultural Resources for the Northeast Region) and Charlie Pepper (Deputy Director of the Olmsted Center).

The trip's discussion centered on the need to continue the work begun in 1992 by Dr. Nora Mitchell of the former North Atlantic Region, in cooperation with Dr. William M. Coli of the University of Massachusetts, Amherst, which resulted in a preliminary inventory of old orchards and fruit trees in the national parks. The service wide inventory revealed approximately one-third of all national park units had old orchards and fruit trees, yet the National Park Service knew very little about the trees and almost no preservation maintenance was being performed.

As a result of that memorable trip I found myself assigned the plum task of preparing a historic context study of orchards in the United States, which would provide the basis for future evaluations of significance and integrity of orchards and fruit trees in the national parks. Once the historic context study was prepared, more in-depth inventories and evaluations could follow, and a program of historic orchard management could build from there. Unfortunately, this document has taken longer to write than any of us imagined. The completion of this phase of work marks just the beginning of a long road to better understand and preserve the historic orchards of the national park system. The need is urgent and much work lies ahead. It is my hope that this document will help stimulate greater momentum in moving forward with this huge task.

I've been honored to work on this project with many invaluable contributors. First of all, my sincere thanks are due to my colleagues Bob Page, Charlie Pepper, and Cathy Gilbert for providing me the opportunity to author this document and for much guidance along the way. Great thanks are also due my long-time supervisor Dr. Stephanie Toothman, Chief of Cultural Resources for the Pacific West Region, who has patiently supported my work on this project along with Randall Biallas, Assistant Director, Cultural Resources Programs in the Washington, D.C. office. My thanks are also due to Cari Goetcheus, Assistant Professor, Clemson University, Kimball Koch, Cultural Landscapes Program Lead, Pacific West Region, Margie Coffin Brown, Landscape Architect, and Jill Cowley, Historical Landscape Architect, for their meticulous reviews of drafts and much encouragement over time.

Research for this document has taken me to some wonderful orchards throughout the national park system. Many thanks are due to the NPS staff that generously gave their time in field visits and are actively involved in the preservation of orchards. Traveling from west to east, my sincere thanks are due to Jesse Kennedy, Cultural Resource Specialist and Laurie Thompson, Orchard Manager of North Cascades National Park Service Complex, for taking great care of the Buckner Orchard, which to a great degree has provided the inspiration for this project. My thanks to David Blackburn, former Chief of Interpretation at John Muir National Historic Site, for introducing me to the historic pear orchard at Muir's gravesite and the historic olive orchard on Mt. Wanda.

Thanks to Tom Leatherman, Superintendent of Manzanar National Historic site and his maintenance staff, Gerry Enes and Bob Clyde, for their efforts to stabilize the park's orchards. My thanks to Paul Chattey, former Branch Chief of Historic Structures and Landscapes at Yosemite National Park and Steve Torgerson, former park Historical Landscape Architect, for introducing me to the remarkable orchards of Yosemite Valley. My thanks to Jeff Pace, former Orchard Manager and Lee Kreutzer, former Cultural Resource Manager in Capitol Reef National Park for showing me the national park system's largest orchard complex in the Fruita Historic District.

My thanks are due to Bob Cherry, Resource Management Specialist of Blue Ridge Parkway and Lucy Lawliss, former Cultural Landscapes Program Manager of the Southeast Region, for introducing me to the orchards of the Moses Cone Memorial Park. Thanks are due to Marla McEnaney, Historical Landscape Architect with the Midwest Region for unveiling the Frank Farm orchard on North Manitou Island, in Sleeping Bear Dunes National Lakeshore. My thanks to Heidi Cope, former Horticulturist with the Olmsted Center for Landscape Preservation, Eugene Gabriel, former Facility Manager and Bob Mackenzie, Gardener at Adams National Historical Park, for showing me the splendid Adams orchard.

My thanks to Charlie Pepper, for taking me to see the astounding 150-year old seedling apple trees at Minute Man National Historical Park and introducing me to the more than 200-year old Roberts Farm apple trees at

Delaware Watergap National Recreation Area in New Jersey. Thanks are also due to Maureen Joseph, National Capital Region Cultural Landscapes Program Manager and her staff, Judith Earley and Jennifer Hanna for showing me the remarkable McIntosh trees at Oxon Hill Farm in D.C. I am also grateful to Anna Tamura, NPS Landscape Architect and Research Assistant during the early phase of the project, for spending many hours with horticultural texts in the University of Washington libraries.

My research has also led to memorable interactions with partners and experts outside of the NPS. My sincere thanks are due to Dr. Kim Hummer, Research Leader and Curator of the United States Department of Agriculture, Agricultural Research Service, National Plant Germplasm Repository in Corvallis, Oregon, and her colleague Joseph Postman, Plant Pathologist, for introducing me to cryogenic germplasm conservation, DNA mapping and the National Plant Germplasm Repository living collections. My thanks to Dr. Bob Norton and Gary Moulton of the Washington State University Mount Vernon Research and Extension Unit, for lessons on contemporary orchard cultivation. My thanks to Dr. Bruce Barritt, Apple Breeding/Orchard Management Systems at Washington State University Tree Fruit Research and Extension Center, for lessons on high density orchard management systems and variety obsolescence. My thanks to Marca Woodhams and her staff at the Smithsonian Institution Horticultural Library for access to historic horticultural texts and catalogs. My sincere thanks to Tom Burford, author and orchard and nursery consultant, for sharing his knowledge of the history of orchards and great orchard wisdom. Last, but by no means least, my deep gratitude is due to Robert Z. Melnick, former Dean of the School of Architecture and Allied Arts at the University of Oregon and mentor in cultural landscape preservation, for demonstrating that an orchard, namely the Dorris Ranch in Springfield, Oregon, could be nominated to the National Register of Historic Places.

INTRODUCTION

The history of orchard fruit growing in the United States is as rich and complex as the country's history of human settlement and development. But the history of orchards and orchard fruit cannot be told as one story. Orchards have evolved along a path characterized by steady, sustained growth followed by dramatically increasing complexity during the past 100 years. In contrast, the evolutionary path of orchard fruits is characterized by a simple beginning followed by the rapid introduction of numerous fruit varieties, followed almost as rapidly by a tremendous loss in diversity. By the late 1800s the histories of orchards and orchard fruits were on divergent paths. Orchards continued to become more complex, but the diversity of fruits began to decline. Throughout the 1900s, orchards were transformed by horticultural science and technology, whereas orchard fruits became so diminished in number that the extinction of thousands of varieties occurred. Why did these changes occur, and how did they manifest themselves in the appearance of orchards in the national parks?

Purpose

The purpose of this historic context study is to provide information about the development of the most common types of orchards and fruit trees in the United States and to encourage the nomination of these potential cultural landscapes to the National Register of Historic Places. This study describes the events, important individuals, patterns of development, and spatial characteristics associated with these horticultural properties. The study also provides a national context for identifying and evaluating landscapes that may be eligible under Criteria A through D as defined by the National Register. While this historic context was prepared primarily for national park managers, the context is national in scope and may be useful to orchard managers and researchers outside the national park system.

This context study addresses the most common types of old orchards in the national parks that were identified by a 1992 inventory. The study emphasizes most woody, temperate fruit trees such as apple, apricot, cherry, olive, peach, and pear, but generally not tropical or non-woody (monocotyledonous) fruit trees, such as banana, pineapple, avocado, and papaya. Some historic context is provided for citrus orchards, such as lemon and orange, and some context is provided for nut orchards, such as almond, hazelnut, pecan and walnut. However, the majority of context is focused on the most common temperate orchards.

Fruit trees and orchards are largely unevaluated but potentially valuable cultural resources within the national parks. A 1992 survey of national parks revealed that 127 units in the system had fruit trees or orchards, encompassing every important fruit and nut type. Many were old trees within historic sites or cultural landscapes that had not yet been evaluated for their historic significance. However, in some national park units, fruit trees or orchards were already known to be associated with the historic significance of a site. For example the peach orchard at Gettysburg National Battlefield is known to have been important in troop maneuvers during the Civil War. However in many parks, fruit trees or orchards are part of a vernacular landscape that was settled prior to the designation of the park, and reflect less dramatic but still important cultural uses of land. This document will enable park managers to relate fruit trees to the overall history of fruit growing in the United States, and will provide a basis for understanding fruit trees as significant features of cultural landscapes, or in the case of orchards, as potentially significant landscapes in their own right.

Background

This historic context study builds upon the work initiated by the National Park Service Washington Office and former North Atlantic Regional Office, in cooperation with the University of Massachusetts, Amherst Extension Service in the late 1980s. The first NPS Historic Orchard Workshop, held at the University of Massachusetts, Amherst in February 1987, identified the lack of information about old orchards and fruit trees in the national parks as a critical impediment to cultural landscape preservation. Of paramount concern was the lack of identification of fruit trees and their constituent varieties, and the corresponding risk of losing genetic resources and diversity when old fruit trees in the national parks died.

With the support of NPS Director William Penn Mott, Jr., a landscape architect with great interest in historic orchards, the North Atlantic Regional Office Cultural Landscapes Program (under the leadership of Dr. Nora J. Mitchell) initiated a service wide project to identify fruit trees in the national parks. The project was conducted over several years, during which time all units of the national park system were surveyed and many parks were visited. To the extent possible, fruit trees were identified by species, variety, and age, and preliminary management recommendations were provided. As a result, the project's principal investigator, Dr. William M. Coli of the University of Massachusetts, Amherst, produced a report titled, Inventory and Conservation of Genetic Resources in the Form of Historically Significant Fruit and Nut Trees in the National Park System

(1992). Among the report's remarkable findings was the identification of approximately one-third of all park units with old orchards and fruit trees (127 units), and the existence of a number of rare and unusual fruit varieties in the national parks.

By the mid-1990s, the NPS Park Cultural Landscapes Program had developed a methodology and computer software system to begin the implementation of a service wide inventory of all cultural landscapes in the national parks. Called the Cultural Landscapes Inventory, the system outlined a method for documenting, identifying, analyzing, and evaluating cultural landscapes as historic properties, eligible for listing on the National Register of Historic Places. Over the next several years, hundreds of rural areas in the national parks were investigated as potential cultural landscapes, with their constituent agricultural lands, orchards, and fruit trees identified as potentially significant characteristics and features. But the need for more specialized information on historic agricultural and horticultural practices was recognized in order to analyze historic patterns on the land, and to accurately evaluate the authenticity or integrity of the characteristics and features that remained.

The NPS Park Cultural Landscapes Program sponsored a historic context study of agricultural landscapes in the national park system in the mid to late 1990s, which yielded an understanding that patterns in American agricultural landscapes had changed dramatically over time. A corresponding evolution in orchards was anticipated, and interest was spawned to develop a historic context study for American orchards. This document is the product of that study, and it demonstrates that in the last 20 years, interest and understanding of historic orchards has changed markedly. Once perceived as mere repositories of rare or unusual fruit varieties, historic orchards are now seen as distinctive, historically significant cultural landscapes in their own right.

This historic context study attempts to demonstrate that both orchards and fruit varieties have changed greatly over time, and that historic orchards in national parks may be the last remaining examples of period orchards in the United States. Today the conservation of genetic diversity is well recognized as important, and the conservation of germplasm is being conducted by the United States Department of Agriculture (USDA). The USDA operates a network of genebank repositories throughout the United States known as the National Plant Germplasm System. Living collections of rare and unusual fruit varieties exist across the country in nurseries and horticultural collections maintained by individuals and organizations; however, the task of preserving historic orchards in situ is rare.

The National Park Service is mandated by the *National Park Service Cultural Resource Management Guideline, NPS-28* (1997), to lead the nation in cultural resources management and historic preservation. The agency is uniquely positioned, therefore, to preserve historic orchards as cultural landscapes. This historic context study aims to show how old orchards in national parks are worthy of preservation because, for one, they cannot exist elsewhere for social, economic, and technological reasons, and two, rather like a period agricultural landscape or a period designed landscape, historic orchards have unique characteristics that reflect the passage of American history.

Four Periods in the History of American Orchards

The evolution of orchards in the United States can be described in four general periods:

- 1600-1800 when European fruit trees were introduced and planted for both subsistence farming and pleasure.
- 1801-1880 when collectors and entrepreneurs developed fruit varieties.
- 1881-1945 when orchard development focused on commercialization, technology, and regionalism.
- 1946-present when orchard production intensified and dwarf trees became commonplace.

The appearance of orchards and orchard fruits has evolved considerably during these periods, each marked by distinctive characteristics and features. The national park system is a unique land repository of orchards that represent each of the four periods. Orchards with integrity that date from the first three periods may be evaluated as historically significant and listed on the National Register of Historic Places as historic sites, historic districts, or as contributing features to sites or districts. As described in Chapter 5, "Evaluating the Significance and Integrity of Historic Orchards and Fruit Trees," National Register Criteria for Evaluation A through D may be applied. Significant orchards should be protected and managed for their cultural resource values, according to *NPS Cultural Resource Management Guideline, NPS-28* (1997).

During the last 400 years, the appearance of American orchards has changed in the size and form of the fruit tree, the species and varieties grown, the spacing and scale of tree plantings, and the ensemble of associated features, such as fencing, windbreaks, roads, irrigation systems, sheds and barns. Most fundamentally however, two critical trends have changed the appearance of orchards over the four periods: the transition from seedling to variety trees, and the change from standard to dwarf rootstocks. In order to understand these trends it is important to comprehend two interrelated concepts of horticulture that apply to orchard cultivation: vegetative propagation and grafting.

For more than 2,500 years orchardists have known that most fruit trees grown from seed do not bear fruits that are good enough to eat raw. But once in a while, either hybridization or a chance genetic mutation produces a seedling fruit tree that can bear exceptionally tasty fruit. This fruit tree is known as a special selection, or a "variety," and it can be kept in perpetuity only through rooted cuttings. Because the seeds taken from the special fruit do not have the same characteristics as the parent, vegetative propagation (cuttings rather than seed) is used to perpetuate the variety. Vegetative propagation is a key principle in the history of orchards.

Orchardists have also long understood another significant fact: cuttings from variety fruit trees are often very reluctant to form their own roots. To encourage growth, cuttings from the variety are joined to the roots of another fruit tree, generally of the same genus or species, to form one new, conjoined rooted tree of the same special variety. This technique, known as grafting, was practiced in ancient China and Rome (Figure 0.1). Over centuries, orchardists discovered that the rooted part of the conjoined tree, known as the rootstock, could confer particular characteristics upon the variety part, or scion. Rootstocks derived from seedling trees (trees grown from seed) would confer great vigor to the scion, or the aerial parts of the variety tree, resulting in a full-sized or standard variety tree. Other rootstocks could stunt or reduce the vigor of the scion, resulting in a miniaturized or dwarfed variety tree. The practice of grafting variety scion wood onto dwarfing rootstocks has been used in Europe for at least 500 years. Dwarf trees were more easily contained within walled spaces to protect the trees from cold temperatures and wind, and could be more easily pruned into an array of shapes, such as the espalier, than standard trees. The use of dwarfing rootstocks in the United States for commercial fruit production is a trend of the late 1900s. The concept that almost all variety fruit trees are composed of at least two trees conjoined—a scion and a rootstock—is another key principle in the history of orchards.

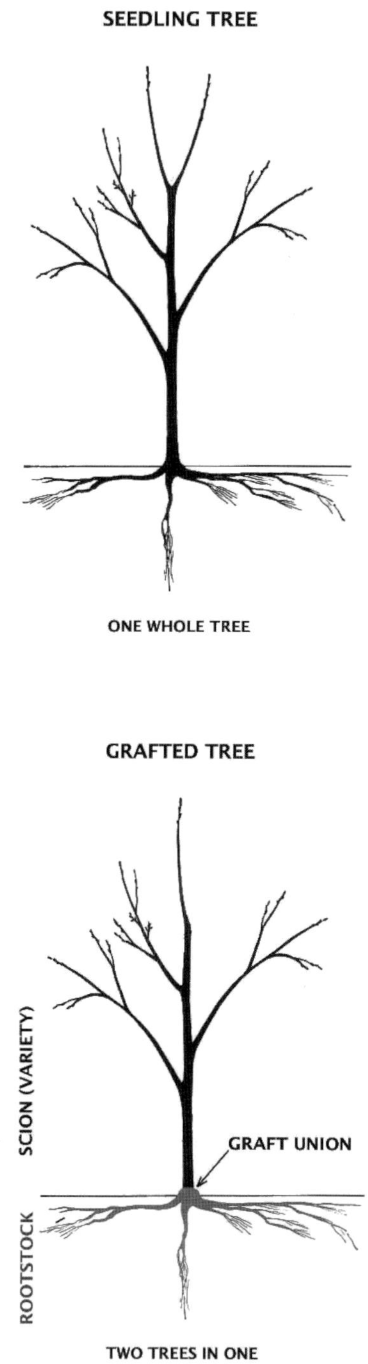

Figure 0.1: Diagram comparing a seedling fruit tree with a grafted fruit tree (S. Dolan).

Throughout the four time periods, the evolution of orchards in the United States is marked by these trends:

- The transition to variety trees rather than seedling trees.
- The transition to dwarf trees rather than full-sized or standard trees.
- An increase in the number of varieties grown during the past several centuries, but a dramatic decrease during the past 100 years.

1600-1800: Fruit Introduction and Colonization

European fruit introductions, subsistence farm orchards, and fruit gardening for pleasure. *For a complete description of this period see Chapter 1.*

During this period orchard fruits were first introduced into America by European settlers, and most orchards were started from seed, rather than from trees planted out. These early orchards were grown principally for cider production and animal feed, rather than for edible table fruits. Orchards were highly irregular in form as a result of sowing, and farm orchards of three to five acres per farm consisted of ungrafted seedling trees of no variety. Seedling fruit trees were highly irregular in form, rather like forest trees, and virtually no pruning was performed. However, livestock was allowed to graze beneath trees, where they browsed off young tree limbs. As a result, seedling fruit trees had tall trunks, often over six feet tall. By contrast, wealthy landowners had fruit gardens, consisting of many European variety fruit trees that were grafted onto dwarfing rootstocks. These trees were highly pruned into various forms within a fenced or walled, ornamental enclosure. Fruit gardens were grown principally to provide edible table fruits. Their owners obtained young, grafted trees through plant nurseries by shipment in barrels from Europe, or by the late 18th century from nurseries established in Massachusetts, New York and Virginia.

1801-1880: Fruit Diversification and Migration

American fruit varieties, collectors, connoisseurs and entrepreneurs. *For more information about this period see Chapter 2.*

During this period, grafting and planting became standard practice for starting an orchard rather than sowing, marking a transition from cider production to edible table fruits of all kinds. Variety trees replaced seedlings because their fruit was good to eat raw, and hundreds and sometimes thousands of American orchard fruit varieties were created by American plantsmen. Commercial orchards were established, first on the East Coast and then throughout the United States. Variety fruit trees were spread throughout the country by migrating settlers. Horticultural literature, horticultural societies and shows popularized American fruit varieties, but still very little pruning or other horticultural management was done to fruit trees. Orchards had a more geometric form due to the use of grafted variety trees rather than sowing seeds, though individual trees still had a relatively natural, unpruned form with a tall trunk as in the previous period. Variety trees were grafted onto seedling rootstocks, giving rise to full-sized, standard trees (Figure 0.2). Despite a transition to variety trees, cider apple orchards were still started from seed rather than from grafted trees.

Figure 0.2: Diagram of the four periods in the history of American orchards (S. Dolan).

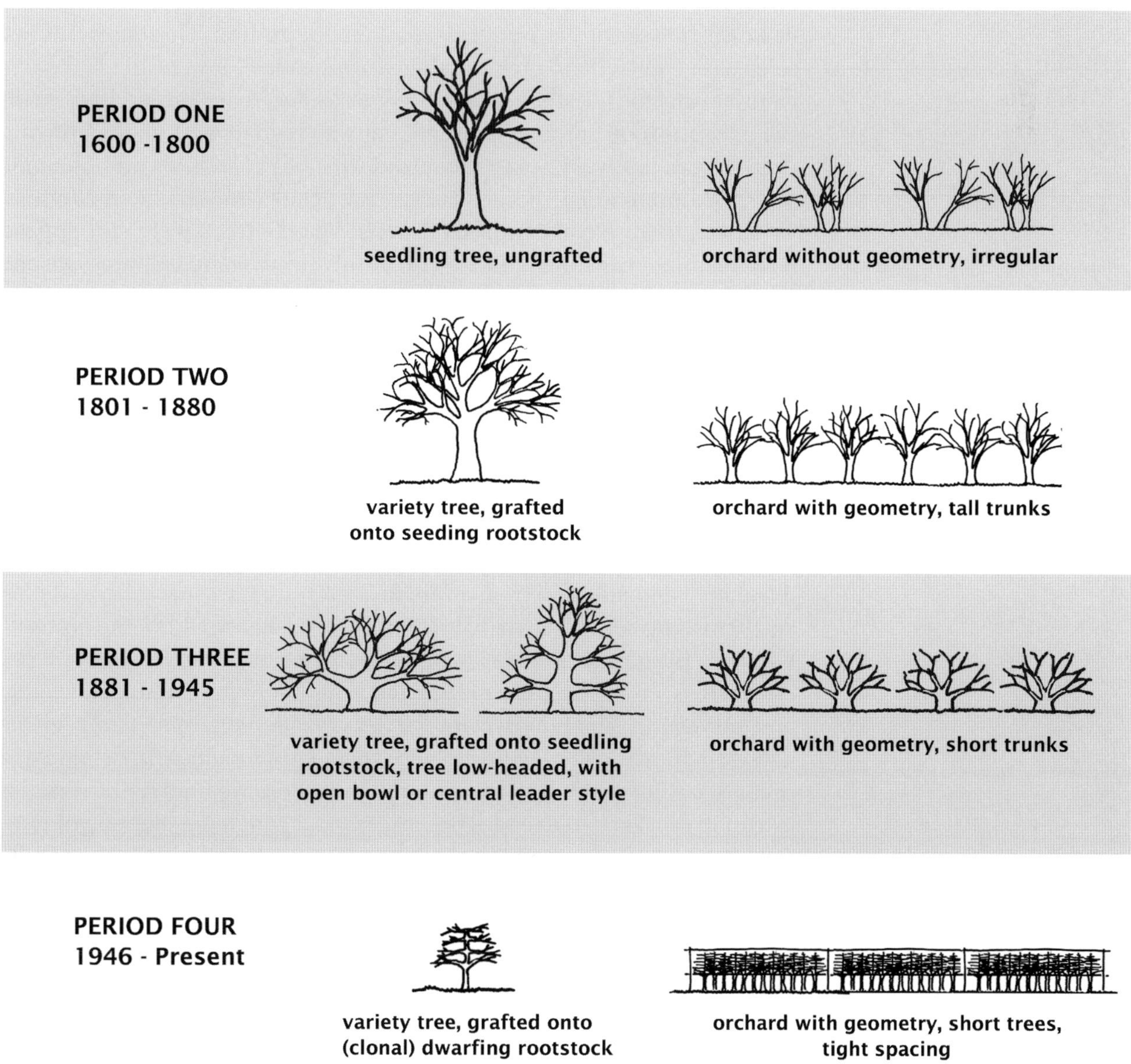

1881-1945: Orchard Specialization and Industrialization

Commercialization, scientific and technological development and regionalism. *For complete information see Chapter 3.*

During this period, horticultural practices became standard for maintaining an orchard. The United States Department of Agriculture, formed in the late 1870s, gave rise to a new generation of scientifically educated orchard growers, using newly invented pesticides, mechanical irrigation systems, cold storage, tractor-powered equipment, and grading systems. Pruning in one of two forms became the norm: either central leader or open bowl form, and tree trunks were dramatically reduced in height to attempt to make the tree canopy more accessible from the ground. The appearance of trees with trunks 18-36 inches in height was called "low-headed." Lower canopies meant livestock was strictly excluded from the orchard, with the exception of poultry. Tree spacing was regular and wide, allowing draft horses and vehicles to move and turn between rows and columns of trees. Commercial characteristics drastically reduced the number of varieties grown from hundreds to tens. Variety trees were still grafted onto seedling rootstocks, giving rise to full-sized, standard trees, despite their short trunks.

1946-Present: Fruit Monoculture and Orchard

Intensification Internationalism, dwarf trees and high density management systems. *For a complete description see Chapter 4.*

During this period American farm orchards disappeared and commercial orchards became highly sophisticated, technology-driven operations. Clonal, or cloned dwarfing rootstocks, rather than seedling rootstocks were developed and adopted by growers for many orchard species, producing smaller trees and dramatically increasing tree density within the orchard. Growers began using trellises or stakes to permanently support weak dwarf trees, and some used mechanical equipment for pruning, thinning, and harvesting. The number of varieties considered commercially viable dropped dramatically, leading to widespread monoculture in commercial orchards. After reaching a low point in the 1980s, the number of varieties grown in the last two decades has increased through the planting of new, rather than old, varieties (with a few exceptions). Today, universities are moving into the role of new variety development and patenting varieties for license to specific growers. Thousands of old fruit varieties are now extinct through lack of propagation and cultivation. The appearance of the pre-World War II orchard has been transformed through the use of short, tightly spaced, and short-lived fruit trees.

Part I:

A Historic Context of Orchards In the United States From 1600 to the Present

CHAPTER 1
FRUIT INTRODUCTION AND COLONIZATION, 1600-1800

European fruit introductions, subsistence farm orchards, and fruit gardening for pleasure

Introductions by Seed

The origins of orchard fruit growing in the United States were in fruit seeds carried from the Old World by sea-weary European settlers and missionaries. These seeds, when sown on land cleared of forest or brush, were among the first non-native plants to transform the landscape of the New World. Their transport to America led to a dramatic evolution of their species, for in the next several hundred years Americans would transform them into thousands of new varieties.

Most orchard fruits that are presently of economic or cultural significance in the United States, such as almond, apple, apricot, cherry, hazelnut, lemon, orange, peach, pear, plum, quince, and walnut, are not native, but were originally brought from Europe by seed. The introduction of peach to Florida in the mid-1500s and the pear, apple, cherry, and plum to the colonies and southern Canada in the early 1600s led to the transformation of orchard fruits in America and their dissemination throughout every part of the country.

Origins of Horticulture

The seeds carried by settlers were of familiar fruits from the old countries where orchard fruits had been improved over thousands of years of domestication. Most orchard fruits sown in 17th century Massachusetts, New York, or Virginia were known to ancient Rome. Romans were instrumental in spreading fruits indigenous to Asia and southern Europe all over the empire, and introducing apples and pears to England during their occupation from 43 to 407 CE (Hedrick 21: 25). Numerous writings of ancient Rome document the cultivation of apple, pear, peach, cherry and plum, including Homer's *Odyssey* in approximately 1000 BCE. Marcus Portius Cato wrote a treatise on farming in 150 BCE in which he described methods of fruit propagation and cultivation similar to modern methods. In the 1st century, Pliny the Elder wrote about fruits and their cultivated varieties in his *Naturalis Historia*.

With the collapse of the Roman Empire, much of Europe and fruit cultivation retreated into the Dark Ages, and fruit varieties known to the Romans were either lost or maintained only by monasteries. Evidence of European fruit cultivation reappeared in literature in the late Middle Ages, when fruit trees were described within monastery cloisters or in the walled gardens of monarchs. As Europeans began to leave for the New World in the 1600s, horticulture was emerging as a pastime of the leisure classes in Europe. The first English horticultural writings were published in the 1500s, in which walled gardens with fruit trees were described as pleasure grounds. These gardens of medieval gentry had lawns with formal arrangements of walks lined with fruit trees. Clipped fruit hedges ran along masonry garden walls, and were occasionally punctuated by turf seats.

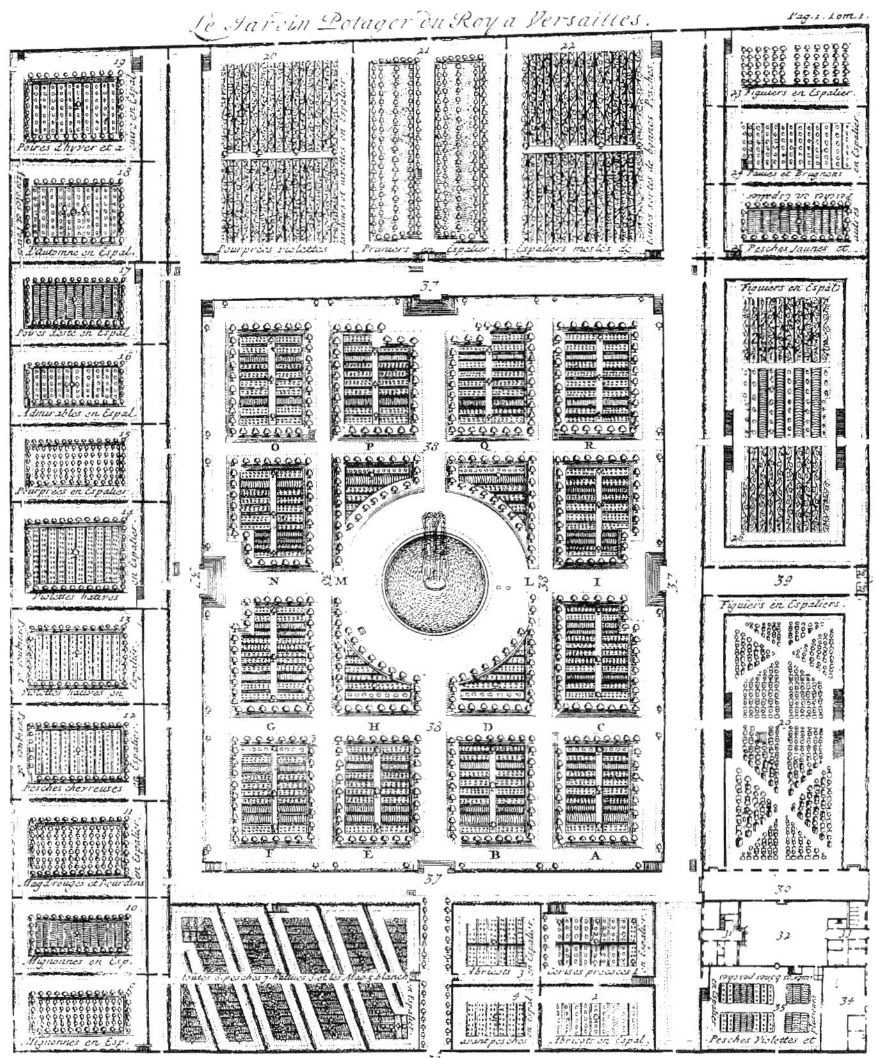

Figure 1.1: Louis XIV fruit garden at Versailles, designed by La Quiutinye (1626-1688) with hundreds of dwarf fruit trees of true varieties (from Tukey, 1964, courtesy of Cornell University Press).

A range of intensively cultivated fruits could be found in these gardens, including quince, apple, pear, cherry, plum, apricot, and peach (Greenoak 83: 10). The fruit gardens of 17th century Versailles contained hundreds of dwarf fruit trees (Figure 1.1).

Fruit Gardens

The intensive cultivation of fruits found a home in America within the walls of the fruit gardens of nobility. Prominent settlers with manorial rights granted by the English monarch, or wealthy landowners with estates, developed walled gardens close to the house, where fruit trees were densely planted in formal arrangements and pruned into an array of hedge and espaliered forms (Figures 1.2 and 1.3). The fruit garden reached its zenith in the 1700s and early 1800s when fashionable English horticultural

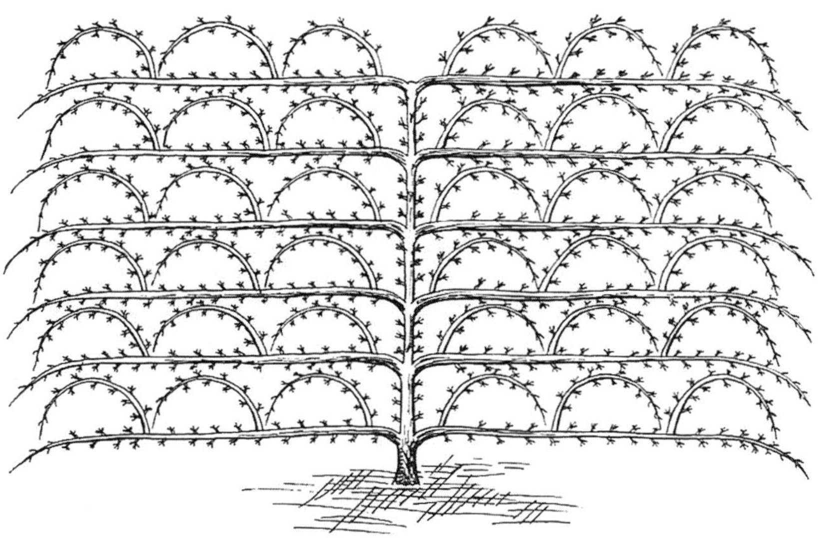

Figure 1.2: Two-dimensional ornamental tree form known as an espalier. This particular shape is known as a "palmette" (from Tukey, 1964, courtesy of Cornell University Press).

Figure 1.3: Dwarf pear trees in a fruit garden at Les Jardins du Luxembourg, Paris, France. This type of pear tree would have been grown in an American fruit garden (S. Dolan, 2004).

books such as Batty Langley's *Pomona: or The Fruit Garden Illustrated*, influenced the selection and intensive cultivation of a wide range of fine European fruits, including apricots, nectarines, and almonds. The fruit garden, a highly controlled, semi-ornamental and productive space tended by gardeners, existed in contrast with another form of early plantation in which fruit trees were grown on American soil: the farm orchard.

Farm Orchards

Beginning in the early 17th century colonies and continuing until the late 19th century, farm orchards were a uniquely American expression of fruit growing. Farm orchards were the plantations of humble farmers, who cleared their land and sowed apple seeds en-mass over their few acres to create an irregular orchard of 250 to 300 apple trees, or approximately 50 trees per acre (Figures 1.4 and 1.5). In the southern colonies, peach seeds were sown separately from apples, with a typical farm orchard of peaches having 1,000 trees, or approximately 150 trees per acre (Hatch 98: 13). Farm orchards were highly distinguishable from the specialized fruit gardens of landed gentry, or the geometric commercial orchards that would later develop in the early 1800s. Farm orchards were sprouted from seed rather than from young trees planted out, and emerged from the soil wherever apple seeds or peach stones were buried. These orchards grew up without a formal geometry, and with great variability in their constituent trees (Figures 1.6 and 1.7). Seedling orchards were also started in informal "nurseries," or where animals had deposited seed. Sometimes these trees were transplanted into seedling orchards with greater regularity to their layout.

Figure 1.4: Children picking apples from seedling trees in an 18th-century farm orchard, artist unknown. Note the tall tree trunks and the "wildness" of the orchard (courtesy of Fedco).

Figure 1.5: Apple pickers in an 18th-century seedling farm orchard, artist unknown. Note the "wildness" of the tree form and the presence of fowl (courtesy of the Illustrated London News, 1873).

Figure 1.6: Oil painting of apple harvest in an 18th century seedling farm orchard. Note the tall tree trunks and the irregular form and spacing of the trees (artist unknown, courtesy of Fenimore Art Museum, Cooperstown, NY).

Figure 1.7: *(below)* Historic woodcut print of 18th-century harvesters in a seedling farm orchard. Note the tall height of the ladders and tree canopies (artist unknown, courtesy of Fedco Trees).

Seedling Fruit Trees

A fruit tree raised from seed, like any flowering plant, is a unique individual, and the size and shape of the tree, and the size, shape, color, and taste of the fruit will differ from one seedling fruit tree to the next. The seedling farm orchard was therefore an irregular collection of trees, each with a unique character, blossoming and fruiting at different times. Despite the great genetic diversity in such orchards, the fruit was rarely palatable raw, and for the first 200 years of settlement in America, the fruits of farm orchards were used almost exclusively for making alcoholic beverages and feeding livestock (Hedrick 33: 225).

Although fruit growing is one of the earliest forms of horticulture in America, the farm orchards of the 1600s and 1700s were not horticultural enterprises by modern standards. Most orchards were uncultivated, unpruned, and unwatered, and fertilizing was performed passively by grazing livestock and poultry. Animals provided manure for the orchard and also aided in pest control by feeding upon insects and fallen fruit with disease residues. Hogs and chickens also helped to keep the orchard floor weed-free by foraging on herbaceous plants beneath the trees. Frequently, hogs and chickens were contained in farm orchards by ditches, fences, or hedges circumventing the plantation. The presence of animals in the farm orchard led to browsing upon the trees and the development of a tall, unbranched trunk. Overall, the seedling orchard appeared as a relatively "wild forest" of trees with no strict geometry. Each tree had a naturalistic grow habit, due to a lack of pruning and very tall trunks, of six feet tall or greater (Figures 1.8, 1.9, 1.10, and 1.11).

Figure 1.8: Historic photograph of a seedling apple tree in the late 1800s with a wild, forest-like form and tall trunk. The tree was 150 years old when photographed and was sown in the mid 1700s, CT (from Lowther 1914, courtesy of New York State College of Agriculture).

Figure 1.9: *(below, left)* Photograph of seedling apple trees at Minute Man National Historical Park in Concord, MA. Note the tall tree trunks and "wild" character of trees (S. Dolan, 2001).

Figure 1.10: *(below, right)* Photograph of more than 200-year-old seedling apple trees at Roberts Farm, DE, Watergap National Recreation Area, NJ (C. Pepper, 2007).

Figure 1.11: Historic photograph taken in 1914 of a seedling apple tree displaying a characteristically tall trunk and irregular form. The tree was sown in 1825 by the Hudson's Bay Company at Fort Vancouver. The tree is still alive and was used in 2004 to propagate seedling apple trees for a reconstruction orchard at Fort Vancouver National Historic Site, WA (Lowther, 1914, courtesy of New York State College of Agriculture).

Role of Cider

In colonial America the most important beverage was cider. The importance can hardly be overstated, as cider was both the beverage of subsistence, before potable water, and a commodity for trade. Before grapes could be successfully grown in America and before ale could be cheaply consumed, hard cider was a staple for all classes of Americans and a form of currency for goods and services. A five-acre farm orchard of seed-grown or seedling apple trees could yield more than 1,000 gallons of cider per year; enough to supply an extended family through the seasons. The arduous cider-making process involved grinding the seedling apples to a pulp through an apple mill, then leaving the pulp to steep and ferment until its color deepened.

The pulp was then put through a press and the collected liquor was strained through a sieve and poured into casks. Filled casks were left unbunged in a cool cellar, where fermentation was allowed to continue for a few days. At the right time, charcoal was added to stop the fermentation, and casks were then bunged, sealing in the alcohol (Downing, 47: 62).

Seedling Apple and Peach

Cider, along with late apples harvested in October and early November, could be kept throughout the winter and spring in cold cellars. Although seedling apples were generally not eaten raw because of their astringent taste, they had many culinary uses and were extensively used in baking and drying. Some seedling fruit trees, including seedling apple trees, did produce pleasantly edible fruits, though in general, the edibility of their fruits was considered very unreliable. Seedling peaches could not be stored over winter like apples, but the fruits were more important seasonally for animal feed and for making brandy. Seedling peach trees had other uses too; the strong and fine-grained wood was used to make furniture and tool handles, and was commonly used as firewood when the short-lived trees reached the end of their productive lives. Peach trees were also used as living fences, and were densely planted to form hedges around southern plantations.

Seedling Pear

As farm orchards of apples and peaches were sown in the early 17th century colonies, pears were being introduced into the New World by the French in Nova Scotia, Cape Breton, Prince Edward Island, and Montreal, Canada. By 1600 the French had done more to domesticate the pear than any country. Pear cultivation in France is known to date to the 9th century, when Charlemagne, ruler of the Franks, had numerous varieties of pears cultivated. In 1608 Olivier De Serres, the French "Father of Agriculture," wrote the first original horticultural text since the Roman Empire, and described numerous varieties of pears, including summer, fall, and winter ripening varieties. In 1628 the French horticultural writer Le Lectier described 254 pear varieties (Hedrick 21: 18).

French immigrants brought fruit seeds to the New World, like English settlers in Massachusetts and Virginia. While the cultivated varieties of pear trees they had left behind in France bore fine dessert fruits, the fruits of their seedling pear trees in the New World were not good to eat raw. French missionaries and settlers sowed pear seeds around their dwellings and established mixed farm orchards of pears and other fruits. French missionaries introduced the pear to the aboriginal peoples of Canada, just as Spanish missionaries were introducing the peach to American Indians in Florida and the Gulf States. The dispersal of these fruits followed the missionaries' migrations into the interior of the continent. By 1629 pear seeds had arrived in Massachusetts from England and were sown within farm orchards and around farmhouses. Pears were sown less prolifically than apples, however, as seedling pears had fewer uses and had to be cooked before consumption. Seedling pears were made into an alcoholic beverage called perry, but this was much less common than cider, and seedling pears were fed to livestock. Seedling pear orchards were found more in the cooler northern climates of the country.

American Indian Cultivation

European settlers were not the first North Americans to grow fruits. In the mid-1500s, when Europeans made their first contact with American Indians, plums were found growing in the villages of native peoples (Hedrick 33, 27). These plums, of which there are many native species, did not become commercially important in the orchards of Euro-Americans until the 1800s when the first native plum varieties were created. (In the mean time, European plums were introduced and planted in orchards by European settlers.) Native plums were important subsistence fruits for American Indians, and were gradually adopted in the farm orchards of European settlers. The Chickasaw plum, ***Prunus angustifolia***, was among the most palatable and widely grown native plum east of the Mississippi. After contact with Europeans, American Indians acquired apple, pear, and peach seeds through trade with settlers and through missionaries, and began to cultivate these non-native fruits.

The mid-1600s gave rise to the beginning of orchard fruit development in America. Within the first 50 years of settlement, Euro-Americans began to improve the quality of fruits they had brought as seeds from the Old World. The fruit gardens of wealthy landowners and gentlemen farmers became the first horticultural experiment stations, as fruit collectors and early connoisseurs began the process of selecting particularly good seedling fruit trees and propagating them vegetatively through cuttings to create clones.

First American Varieties

By the 1650s, the first American varieties of apples had been created by selection from seedling trees, and among the first varieties were two that would have great commercial importance until the 20th century. Roxbury Russet, an apple with a brown, russetted skin and a rich acid flavor, was created within the developing city of Boston in Massachusetts. Rhode Island Greening, an apple with a pale yellow-green skin and a brisk acid flavor, was another early American variety created in the colony of Rhode Island. Roxbury Russet and Rhode Island Greening were unusual for their time in that their popularity spread beyond the vicinity of their development and became regionally important for cider and as good dessert fruits.

First Imported Varieties

In the mid-1600s, another trend was beginning that would gradually change orchard fruits in the New World: a trend towards the use of young trees for planting orchards, rather than starting by seed. The first importation of grafted fruit trees was recorded in 1647 when the Dutch governor of New York acquired a pear tree from France. The young pear tree was the Summer Bon Chretien variety, possibly the progenitor of the Bartlett variety, and it was shipped from Holland across the Atlantic in a barrel and presented to Governor Peter Stuyvesant at his mansion house in New Amsterdam.

Governor Stuyvesant was an avid fruit collector and is known to have developed a large and highly varied fruit garden named "The Bowery." Planted in the governor's fruit garden and tended by slaves, the imported pear tree lived for the next 200 years as the city streets of the newly named New York City grew up around it, and the Revolutionary and Civil Wars came and went. The huge old tree was destroyed in 1866 when a runaway dray, or delivery wagon, collided with it at the corner of Third Avenue and Thirteenth Street (Beach 05: I, 18). The genes of the early French pear lived on, however, in the many cuttings taken from the tree. These cuttings were distributed among Dutch settlers along the Hudson Valley. Commercial pear growing, which still exists in the Hudson Valley today, can be traced to the 17th century Dutch settlers who had access to trees of French varieties.

Seedling Tree Longevity

Although Governor Stuyvesant's pear was the first recorded importation of a foreign fruit tree, an even older imported pear is known to have been grown on American soil from 1632—the Endicott pear. Owned by Governor John Endicott of Massachusetts, this was a seedling tree rather than a variety and it's unknown whether the tree was started from an imported tree or from an imported seed. Perhaps more remarkable than the recorded history of the tree is the fact that the tree is recorded as presently alive, making it more than 375 years old! Growing on Endicott Street in Danvers, Massachusctts, this pear tree has survived damage by hurricanes in 1804, 1815, 1843, and 1934, and a severe attack of vandalism in 1964. After each round of damage, the seedling tree re-sprouted from the trunk and re-grew. In 1997 the tree's germplasm was conserved in perpetuity by its addition to the pear collection of the National Plant Germplasm Repository in Corvallis, Oregon (Postman, 02: n.p).

Beginnings of Grafting

Once the first American apple varieties were developed and the European varieties of fruit trees were introduced, the practice of fruit tree grafting began in America. The technique of grafting a shoot of one young fruit tree of a known variety onto the roots of a wild and vigorous seedling

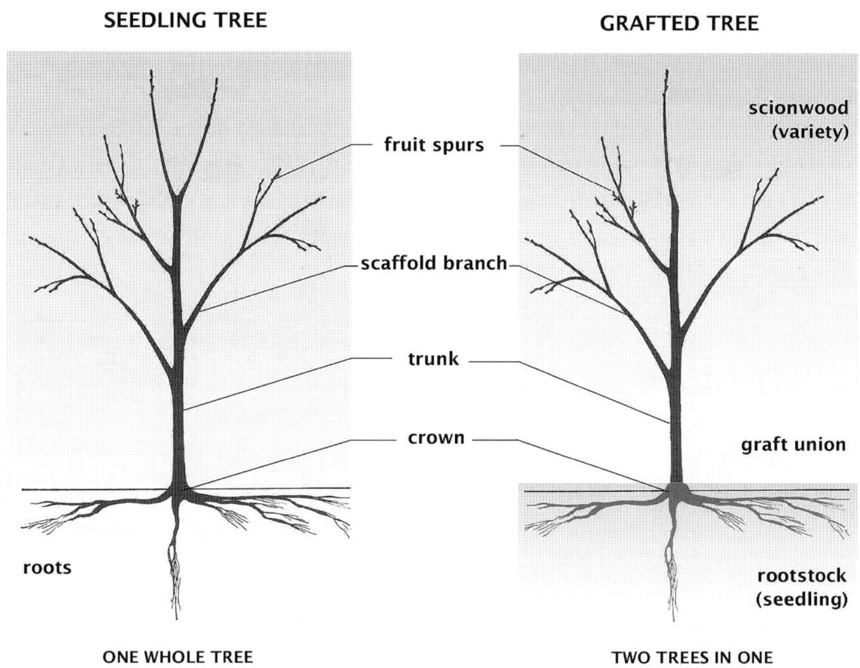

Figure 1.12: Diagram contrasting the form of a seedling tree with a grafted tree (S. Dolan).

tree had been performed in ancient Rome and China. However, it wasn't until cultivated varieties were available in the mid-17th century that grafting began in America (Beach 05: I, 25). Throughout the late 1600s and much of the 1700s, grafting was practiced to a very limited extent by educated fruit growers and some early nurseries, such as the Ralls Nursery in Virginia.

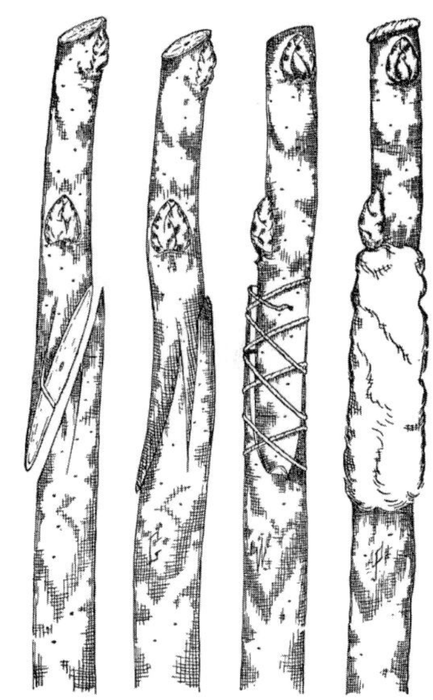

Figure 1.13: Diagram showing the steps involved in grafting scionwood to a rootstock (from Lowther 1929, courtesy of New York State College of Agriculture).

Fruit collectors with fruit gardens and leisure time were the first nurserymen. They propagated European varieties of pear, plum, peach, and later on apricot, nectarine, and fig from young imported trees by taking cuttings and grafting them onto the roots of fruit trees grown from seed in America. The combination of the shoot from the tree that would produce good fruit—the scion—in union with the vigorous roots of a tree grown from seed—the rootstock—was known to produce a clone of the fruit variety with more hardiness and vigor than an ungrafted tree (Figures 1.12 and 1.13).

Grafting was, however, regarded with suspicion by the majority of farmers in America, and most farm orchards remained composed of seedling trees rather than variety trees for the next 150 years. One reason for the American farmers' lack of interest in grafting and growing fruit varieties was the suitability of their seedling fruit trees for cider and animal feed. The diverse characteristics of seedling trees worked well for cider production, as a blend of many different apples produced good flavor. Additionally, the continuous, non-synchronous nature of seedling fruit bearing extended the season of harvest. Cider was equally important in the farm economy of England, though farm orchards were smaller in size than in the United States. In England cider served as wages and was used for barter and sustenance. It has been suggested that the popularity of cider in England may have also impeded the development of grafted apple varieties there. Only in countries where dessert apples were more important than cider, such as in France, did farmers have a compelling reason to develop and graft true varieties (Rosenstein 96: 159).

In the last two decades of the 17th century, French Huguenots settled New Rochelle in Westchester County, New York, and brought with them the pear varieties Doyenne, Brown Beurre, St. Germain, Virgouleuse, Winter Bon Chretien, Bergamot, and Jargonelle as young trees from France (Hedrick 21:49). Quakers from England settled in Philadelphia and sowed all kinds of hardy fruits, including apples, pears, peaches, cherries, and plums. As the century closed, French settlers that had colonized southern Canada in the early 17th century were spreading into the upper Midwest, including Michigan and Illinois. These settlers sowed pear seeds around their farmhouses and among the apple trees in their farm orchards. The pear trees that emerged shared similar characteristics due to lesser genetic variability among pear seedlings than apple.

Seedling Tree Vigor

The majority of pear trees sown along the river valleys of the upper Midwest by French settlers appear to have been related to the variety Doyenne. This is a medium-sized turbinate or bulbous pear, with a clear, lemon-yellow skin. More remarkable than the seedling fruits, which were not so fine quality as the true variety, were the vigor, productivity, healthfulness and longevity of the trees. These trees grew up to 80 feet in height, with a 10-foot circumference trunk. They bore fruit heavily and were immune to the disease, fire blight. Two hundred years later, at the end of the 1800s, these distinctive trees could still be found growing along the Detroit River (Hedrick 21: 52).

Naturalized Seedlings

By 1700 some of the European orchard fruits brought to the New World as seeds had escaped from farm orchards and fruit gardens and began to naturalize. Both peach and cherry spread prolifically throughout the colonies, with peach naturalizing throughout the mid-Atlantic and southern colonies, and cherry more abundant in the north. William Penn observed naturalized peaches around the outskirts of Philadelphia in 1683, and from Pennsylvania through the South, the peach was regarded as a weed. Spread through animal feces and easily sprouted from the stone, seedling peaches flourished along early wagon roads and forest edges. One writer observed that during the 1700s, when a pine forest was cleared in Virginia, a peach grove would generate in its place (Hatch 98: 82). The peach tree was such a common sight that the early American botanist John Bartram mistakenly classified the peach as a native plant.

Seedling Cherry

Naturalized cherries were almost as common in the northern states as seedling peaches were in the South, where they were better adapted to cooler temperatures and were spread by birds. Cherries were introduced in the early 1600s and were possibly the first ornamental plants in America. They were used to decorate the yard around the farmhouse, to line the drive of a wealthy landowner, or to make a border around a fruit garden. During the 1600s, cherries were brought from Europe in the seeds of two species: *Prunus avium*, the Mazzard or sweet cherry, and *Prunus cerasus*, the Cherry plum or sour cherry. Scions or young shoots of choice English varieties were imported by wealthy gentry, and fine grafted specimens such as May Duke and Carnation were planted in the fruit garden. However, peasant farmers regarded sweet and sour cherries as a luxury because they had limited culinary uses and weren't used for cider or livestock feed. Consequently, cherries were rarely found en mass in farm orchards. Instead, one or two cherries around the farmhouse provided a dried fruit treat rather like candy, and plenty of early summer fruit for pies and preserves.

Seedling Plum

Damson plums and quince were fruits commonly found around the farmhouse in the early 1700s, along with sweet and sour cherries. Damson is the name for both the European plum species *Prunus institia*, and its cultivated variety, *Prunus institia* 'Damson.' The Damson grown in the farmyard was not the choice variety, but the vigorous and adaptable seedling tree of the European species produced good culinary fruits. This plum was named for the ancient city of Damascus, and by the time it appeared in the American farmer's yard, it was little altered in more than 2,000 years.

Seedling Quince

Quince was a popular farmyard fruit tree in the northern colonies during the 1700s, but it diminished in importance over the next hundred years. Quince *(Cydonia oblonga)* is an ancient fruit like the Damson plum. Named after a city on the Isle of Crete, it was mentioned by ancient writers, including Pliny. Quince was the standard medieval fruit in Europe, where it was valued for pies and preserves (Greenoak 83: 22). Quince is an adaptable plant that grows as a small, crooked, often multi-trunked tree or large shrub with a broad, rounded crown. It thrived in places where other fruit trees would not grow, such as boggy patches of poorly drained soil and at fence corners. Today the quince is relatively unknown, its demise is due to the unpalatability of the fruit when eaten raw and its high susceptibility to fire blight. Described as a hybrid in taste between orange, apple, and pear, quince must be cooked before eating. By the 1800s interest in preserving the fruit had declined along with the rising popularity of good dessert fruits that could be eaten raw. The fragrant but lowly quince was the first European fruit introduced to America to become relatively forgotten.

First Commercial Nursery

The most important event in the history of fruit growing in America in the first half of the 18th century was the establishment of the first commercial nursery. This made fruit varieties more widely available. Robert Prince, a Long Island landowner associated with the French Huguenots, established a commercial nursery in Flushing in 1730. The Prince Nursery was the only one of its kind at the time and a huge commercial success. The nursery imported cultivated varieties of fruit trees from Europe, propagated them by grafting, and sold them to middle class and wealthy landowners, including George Washington and Thomas Jefferson. The nursery served as a conduit for the introduction of many European fruit varieties into America, and the grafted trees were rapidly planted in fruit gardens throughout the colonies, replacing inferior seedling trees.

The Prince Nursery was also highly significant for the creation of many new fruit varieties in America. The Princes became expert breeders as well as fruit growers, and several generations of Princes created new American varieties that would become important commercial crops. The Prince Nursery was the country's preeminent nursery in the 1700s early 1800s, until the 1860s when the commercial enterprise was closed and the vast plant collection was developed into the Linnaean Botanic Gardens.

European Cherry Varieties

Twenty years after the opening of the Prince Nursery, European varieties of cherry were first offered for sale in America. Cherries were among the first European varieties of fruit to be distributed in the New World through the Prince Nursery, long before European peach varieties were offered. However, unlike peach and apple, cherry would be relatively unchanged by America, as few American varieties would be developed. The cherry species *Prunus avium* and *P. cerasus* are more genetically stable than peach or apple, and have a much lower tendency to hybridize or mutate to become new varieties. The European varieties offered by the Prince Nursery in 1750, such as Carnation, May Duke, Morello, Kentish, Ox Heart, Black Heart, and Bleeding Heart, remained important through the 20th century.

Early American Apple Varieties

As European cherry varieties were made commercially available in the 1750s, two more important American varieties of apple were introduced. Baldwin, a yellow-red-skinned apple with an acid flavor, was developed in Massachusetts, and Newtown Pippin (also known as Newton Pippin, Yellow, Green, or Albemarle Pippin), a green-yellow-skinned apple with a brisk acid flavor, was developed in New York. These early American apple varieties are among the most significant ever created.

Baldwin is a winter apple well adapted to Massachusetts growing conditions. Because it could be harvested in October, it could be stored throughout the winter and spring. Its popularity soon spread beyond the bounds of the state, and during the next 50 years it became the most widely planted apple variety in New England. By the mid-1800s Baldwin was the most important commercial apple variety in New England, and it held this status until the early 20th century when its lack of cold hardiness and tendency to produce fruit biennially hampered its commercial viability. Harvested even later than the Baldwin is the Newtown Pippin, which could be picked in November. Its excellent storage qualities may have lent the apple its legendary status among American apple varieties. Newtown Pippin was the first American apple variety to be exported, which established credibility abroad for American fruit growing. The circumstances behind Newtown Pippin's exportation were unique, and preceded the transatlantic export of other American apple varieties by more than half a century.

In 1759 when Benjamin Franklin was serving as an American diplomat in London, he requested the shipment of several barrels of Newtown Pippin apples from New York. Franklin may have suspected that the complex taste of the apple would remind the English of their own Pippin varieties and be readily accepted, or perhaps the apple's good keeping qualities influenced his selection; but either way, the Newtown Pippins completed the voyage in good condition and were well received in London. This led to the regular shipment of Newtown Pippin apples to England, and also to requests for scion-wood from the trees for grafting.

First Apple Exports

Newtown Pippin was the first New World fruit variety to be introduced into the Old World, where it was planted in the testing grounds of the London Horticultural Society (later to become the Royal Horticultural Society). However, a shorter growing season and different soil type led to the Newtown Pippin being inferior in quality when grown in Britain, fueling the demand for American imports. Newtown Pippin's popularity in England expanded in the 1800s when it led the way for the export of many other American apples. Newtown Pippin orchards in New York were among the first commercial orchards in America. These orchards, north of New York City, were planted almost exclusively for transatlantic shipment of their fruit to England.

European Pear Varieties

In 1767, the Prince Nursery offered European varieties of pear for sale for the first time. The nursery had almost exclusively wealthy customers, who acquired choice pears from France for their fruit gardens. Unlike seedling pears, however, pear varieties did not become a feature of farm orchards. During the late 1700s, peasant farmers continued to sow farm orchards from seed, and apple was still by far the most common fruit grown. Peach was the only other farm orchard fruit of significant acreage. As the Prince Nursery offered European pears for the first time, Thomas Jefferson was beginning to plant fruit trees at his mountaintop estate, Monticello, in Virginia. Jefferson was an avid fruit grower and collector, and a regular customer of the Prince Nursery.

Jefferson's Fruit Garden and Farm Orchard

Between 1767 and 1814, Thomas Jefferson developed a six-acre fruit garden or fruitery, as he called it, and a 400-acre farm orchard, along with two vineyards, two nurseries, and various berry-fruit growing areas (Hatch 98: 7). Jefferson's estate exemplified the two forms of orchards in colonial America: the fruit garden and the farm orchard. When fully developed, the fruit garden (which included the South Orchard,) contained more than 170 varieties of fruit trees including apple, pear, peach, plum, cherry, nectarine, apricot, quince, and almond (Figure 1.14). Most fruits were European varieties, though Jefferson did acquire some American apple varieties. Through the Prince Nursery and his personal contacts abroad, Jefferson amassed a large collection of fruits from other countries, including Italy, France, Belgium and England. In 1778, Jefferson documented his fruit garden in a plan drawing, and kept meticulous notes and observations of his fruit cultivation. The 1778 plan is the most detailed record of an 18th-century orchard in America. It indicates the highly controlled yet ornamental design of the fruit garden. Fruits were separated by species; i.e., apples from pears and peaches from plums, and were spaced on a geometric grid that varied by species. Apple trees were planted on a 25 x 40-foot grid, peach trees at 20 x 25 feet, and other species at 25 x 25 feet.

Figure 1.14: Contemporary aerial photograph and perspective drawing of Thomas Jefferson's 18th-century fruit garden at Monticello, including the south orchard (drawing by Lucia Stanton, courtesy of the Thomas Jefferson Foundation, Monticello).

In contrast to Jefferson's fruit garden, which was tended by European gardeners and slaves, the 400-acre farm orchard was untended, had no strict geometry, and contained only two species of fruit: apple and peach. The apple and peach trees were of no known varieties, as the trees were started from seed. This working orchard had a dual utilitarian purpose: the apples were for cider, and the peaches were for livestock feed.

European Plum and Apple Varieties

Jefferson was very familiar with the English horticultural writers of his day. He was influenced by Batty Langley's *Pomona*, who's recommended European varieties for the fruit garden were made available through the Prince Nursery and other nurseries emerging in the eastern cities. Jefferson also acquired Philip Miller's *The Gardener's Dictionary*, published in 1768, which advocated the growing of fruit in the fruit garden rather than the orchard, as it would provide both fruit and pleasure. By 1771, Jefferson and other gentleman farmers of his class could choose from 31 European varieties of plums from the Prince Nursery, including the celebrated Green Gage, and many European varieties of apple. Sixteen years earlier, a Surrey County, Virginia nursery had advertized a dozen European apple varieties for sale in the *Virginia Gazette* (Burford 07: n.p.).

Early Apple Development

That same year, in 1771, only five American varieties of fruit were available for sale, indicating the nascent stage of American fruit development. All five varieties were apples, and of these only two would become significant varieties: Newtown Pippin and Esopus Spitzenburg. Recently developed by a Dutch farmer named Spitzenburg at Esopus in the Hudson Valley, the Spitzenburg was a yellow-red apple with an acid flavor. The Spitzenburg, like the Newtown Pippin, rose to considerable economic importance in the 19th century, mostly due to its large size, attractive color and exceptional flavor. Jefferson was among the first settlers in Virginia to plant both of these varieties and is credited with introducing and naming the variety Ralls Genet. Ralls Genet was a popular apple in Virginia in the 1800s, though its fame is attributed more to its offspring Fuji, a variety developed in Japan in 1962.

Beginnings of Commercial Horticulture

When the Revolutionary War erupted in 1775, America was a land of 13 English colonies populated by merely 2.5 million people of mostly English heritage, and by many more millions of American Indians, whose populations were rapidly declining. Most settlers were peasant farmers with little more than a few acres of land and a lifestyle that revolved around their farm crops and local hamlet. Farmers living close to a city would haul some of their produce to market by horse and cart for sale or barter, but most of their yield stayed on the farm for home consumption. Only landed gentry had access to a broader network of goods and services, as their zone of influence and opportunity extended beyond their own colony and they could acquire goods from England. Transatlantic commerce did not exist before the Revolutionary War, as the colonies had no leverage to sell goods for profit to their colonial ruler. After the war, the great potential for export trade with England was stymied by further hostilities, manifested in President Jefferson's embargo on international trade in 1809, and the outbreak of the War of 1812. More than a generation would pass before free trade opened up between America and England, feeding the development of the first commercial orchards.

Spanish Fruit Introductions

As the Revolutionary War raged in America, several thousand miles to the west Spanish monks were beginning to transform land through the introduction of exotic fruits. In the western part of what would become Texas, in Arizona, New Mexico and throughout California, Spanish monks were developing missions with extensive farm orchards and vineyards. However, most of the fruits introduced by the Spanish were different to the hardy fruits introduced into colonial America. Generally these fruits were not hardy in temperate climates, but were often Mediterranean or sub-tropical fruits, requiring hot summers and mild winters to thrive. Spanish monks brought figs, olives, oranges, date palms, pomegranates and grapes to the New World as seeds, and cultivated them intensively for the next 50 years. They also brought the seeds of some temperate fruits, such as apple, peach, and pear.

As the Spanish arrived in the New World they encountered extensive irrigation systems created by native peoples. These early irrigation systems of the New World, hand-dug diversion channels and ditch systems that inundated fields with river water, were possibly contemporary with the irrigation systems of ancient civilizations along the Nile, Euphrates, and Tigris Rivers. Spanish monks adopted these methods of irrigation for their mission orchards, and used thousands of native peoples to work the land. Mission orchards were composed of apples, pears, peaches, olives, and oranges, though seedling oranges were by far the most extensive fruit grown. These fruits were noted by the English naval officer Captain George Vancouver in 1792, upon his arrival at the Santa Clara and San Buena Ventura missions.

Seedling Orange

The orange's journey to the New World was via a longer route than the other fruits, however. Citrus fruits are native to western China and northern India, and were introduced into Spain by Moguls in the early 13th century. The seeds brought to the New World were the distant offspring of these Mogul fruits. The Spanish seedling orange became naturalized in California, and was an important food source for the next hundred years, until superseded by the Navel Orange, the first successful orange variety in the United States (Lowther 14: III, 1475).

More Early American Apple Varieties

In 1776 the United States of America declared independence from its colonial ruler, England, and a mercantile economy began to strengthen. Commercial plant nurseries emerged in every state and with improvement in availability, a few cultivated varieties of grafted fruit trees were planted in farm orchards for the first time. The 1780s saw an increase in the creation of North American varieties of apple, with three varieties becoming important beyond their centers of origin: Yellow Belleflower, Westfield Seek-No-Further, and Fameuse. Developed in New Jersey, Yellow Belleflower is a winter apple with a pale yellow blushed skin. It became one of the most important commercial varieties in the Northeast during the 1800s, though it is probably even more renowned for its association with the discovery of the (Red) Delicious variety in the late 1800s. The appearance of the variety Delicious as a chance seedling tree in an Iowa orchard of Yellow Belleflower was one of the most important discoveries in the history of apple development in the United States, and will be described in chapter three. The curiously named Westfield Seek-No-Further was one of the only important apple varieties to be developed in Connecticut. Fameuse, a Canadian apple, had its ancestry in the seedling apples of French settlers along the St. Lawrence River.

First American Plum Varieties

During the 1790s, American varieties of plums were created and offered for sale for the first time. William Prince of the Prince Nursery, who had been experimenting with the Green Gage, the European standard of excellence in plums, developed several new varieties, named White Gage, Red Gage, and Prince's Gage. The 1790s also saw the European apricot variety Moor Park for sale at the Prince Nursery for the first time, a variety that is still considered to be one of the best apricots. European nectarine and peach varieties were also available for the first time. The best nectarine varieties were considered to be Red Roman and Yellow Roman, and among the best peaches were Breast of Venus (Poppa di Venere), Vaga Loggia, and Green Nutmeg, all no longer available in America.

First American Peach Variety

Coincidental with the commercial introduction of European varieties of peach was the creation of the first American peach variety, Heath Cling, and its advertisement for sale at the Prince Nursery. Heath Cling was not only the first-named American peach, it would become the most acclaimed variety, and would be sold universally throughout the 1800s. One legend of the origin of the Heath Cling places the peach on Long Island, where William Prince, Sr. is thought to have found a naturalized seedling tree growing on a heath. The name Cling is derived from the nature of the fruit, of which the peach has two types, the clingstone and the freestone. The types are distinguished by whether the flesh clings to the pit. Heath Cling, other peach varieties and choice plums, nectarines and apricots were all available for the fruit gardens of the wealthy as the 18th century closed.

Summary

The 1600 to 1800 period of fruit introduction and colonization in America centered on the 13 original colonies and the Spanish missions of the Southwest. Fruit introductions were primarily from seed, and throughout this 200-year period the most common orchards were three- to five-acre farm orchards of seedling apples with approximately 50 trees per acre. Farm orchards lacked a regular geometry, as they were sown, rather than planted out, and consisted of wild-looking, unpruned trees, with very tall trunks, greater than six feet. Lower limbs were browsed off by livestock or wildlife. Apples were grown primarily for cider, while peaches, commonly grown in farm orchards of the southern colonies, were used for animal feed.

A small number of wealthy settlers had fruit gardens. These were small, enclosed ornamental spaces in which fruit trees were intensively grown and highly managed. By the late 18th century, numerous European varieties of fruits were available for sale through commercial nurseries, the most important being the Prince Nursery of Long Island, New York. European varieties were planted in fruit gardens as grafted trees, where they produced fine dessert fruit.

The first fruit varieties developed on American soil were of apples from seedling farm orchards, and then later in the fruit gardens of wealthy fruit collectors. Later in the 18th century, American varieties began to be developed by commercial nurseries. By the end of the 18th century, fruit gardens had reached their zenith in fashionability and sophistication, and some farm orchards were beginning to contain one or more grafted trees of apple, peach, or plum so that their fruits could reliably be enjoyed as raw, fresh fruit.

As the 18th century closed, seedling apple, cherry, and plum had been dispersed throughout New England and to the upper Midwest, and apple and peach had been dispersed throughout the mid-Atlantic and southeastern states. Pear was grown predominantly on a limited scale in farmyards throughout the northeastern states, but could be found on a large scale in orchards of French varieties in the Hudson Valley and on

Long Island in New York. Seedling oranges, olives, figs, and grapes had been dispersed throughout present-day California and western Texas by the work of Spanish monks, and American Indians had both introduced European settlers to native plum species, and received many exotic fruits from the Old World. By 1800 the United States of America had been seeded with a great gene pool of European and Asian fruits, and this pool would give rise to the golden age of fruit growing in the 19th century.

Chapter 2 Fruit Diversification and Migration, 1801-1880

American fruit varieties, collectors, connoisseurs and entrepreneurs

The "Golden Age"

The 1800s are known as the "golden age of pomology," the period in which pomology— the science and practice of fruit growing— produced hundreds of new and diverse varieties of fruit. In this period, cultivated varieties of orchard fruits were spread to every state of the expanded union, and commercial orchards were developed in every part of the country. The period from 1801 to 1880 was also the preindustrial age of American orchards. It was the period before science and technology entered the orchard, when "orcharding" became an acceptable way to earn a living, and when orchardists could be found hand-washing their trees with soapy water to keep diseases at bay. Some of the most long-lived orchard trees in the national park system date to this period, such as apple, pear, orange, walnut, pecan and olive. These trees symbolize the period in which American orchard fruits helped build a sense of national identity and were cherished for their unique taste and appearance.

Early 1800s Status

At the outset of the 19th century, most American orchards in the land area east of the Mississippi River were farm orchards. They consisted mainly of apple trees of unknown variety, grown from seed rather than grafted (Figure 2.1). Fruits that could be eaten raw were uncommon in the diets of ordinary citizens, as only those fruits from cultivated varieties were generally good to eat as fresh fruit. For the most part, only wealthy landowners grew true varieties of fruits in fruit gardens, and the vast majority of varieties were of European origin. Very few Americans were trained in horticulture and Europe was still the center of horticultural art and science. However, the United States leapt into a more diversified period of fruit growing in the first decade of the new century with the development of horticultural organizations and the publication of the first American horticultural literature.

Figure 2.1: Historic photograph of a seedling apple tree displaying a characteristically tall trunk and very large size. The tree was 100 years old when photographed, Beckwith, NY (from Lowther, 1914, courtesy of the Encyclopedia of Horticulture Corporation).

Advent of Horticultural Societies and Literature

In the early 1800s, societies were formed throughout New England and in many of the original colonies to promote agriculture and horticulture and disseminate much-needed information to amateur growers. The Massachusetts Horticultural Society and the Albany Horticultural Society were among the most influential early organizations. The first horticultural book for an American audience was published in 1803 when William Cobbett, an American gentleman farmer, revised and edited the most important English horticultural book of the time. Cobbett took William Forsyth's 1802 book, *Treatise on the Culture and Management of Fruit Trees*, the most widely read book of fruit growing in England, and presented it to Americans as *An Epitome of Mr. Forsyth's Treatise on the Culture and Management of Fruit Trees*. Cobbett criticized his fellow fruit growers, saying American orchardists were slovenly and "...they plant and they neglect...," but despite this disparagement, the book was very popular and influenced the growing of cultivated varieties in America.

Bernard McMahon's *American Gardener's Calendar* of 1806 was the first original American horticultural work. Like Cobbett's book, McMahon's *Calendar* emphasized the neglected and unpruned nature of farm orchards, and provided the first rudimentary guide to the care of fruit trees in America. He encouraged farmers to let hogs and poultry manure their orchards, and to stake and prune young fruit trees. McMahon's and Cobbett's books portrayed an image of fruit growing as a serious and skilled enterprise on American soil. While farmers were slow to adopt the authors' laborious suggestions for the treatment of orchards, the book stimulated a widespread interest in fruit varieties.

Commercial Fruit Characteristics

A sharp increase in the rate of development of American apple and peach varieties came with the turn of the 19th century. While the overwhelming majority of varieties offered by nurseries would remain European until the 1830s, a considerable number of American varieties were appearing and some would later have great commercial significance. Foremost among the new apple varieties was Ben Davis, a yellow-red striped apple developed in Virginia at the turn of the century (Figure 2.2). Despite widespread criticism of its thick skin and faltering flavor, during the 19th century the variety rose to prominence as the most widely grown commercial apple between the 32nd and 42nd parallels. One hundred years after its development, Ben Davis was grown from coast to coast, generally south of the Mason-Dixon Line.

Figure 2.2: Watercolor paintings of Ben Davis (left) and King of Tompkins County (right) apple varieties (from Beach, 1905 courtesy of New York State Experiment Station).

Figure 2.3: Watercolor paintings of Winesap apple variety (from Beach, 1905 courtesy of New York State Experiment Station).

The Ben Davis apple set an early precedent for the qualities sought in a commercial apple variety. Even before commercial orchards were legitimate enterprises, Ben Davis inspired many farmers to rip out their seedling trees and plant a field of the new variety. This occurred for several reasons. First, Ben Davis was well adapted to the growing conditions of the warmer regions where cold winters could occur, a quality that was lacking in many European varieties or seedling apples. Second, it bore fruit early in life, a valuable characteristic when apple trees of the time could take 12 years to produce their first crop. Third, Ben Davis bloomed later in the spring than most of its seedling counterparts, avoiding frosts that would ordinarily kill blossoms and reduce the crop yield. Finally, and most importantly, its particularly thick skin resisted bruising, allowing the apple to be transported in barrels and present a good appearance at journey's end. The frequently documented criticism of the palatability of the Ben Davis apple is at odds with the apple's widespread planting. This implies that Ben Davis was ahead of its time in having commercial qualities that ranked higher than taste, a phenomenon not encountered in the selection of varieties until the 1900s.

Apple Variety Diversity

King of Tompkins County, Northern Spy, Winesap, McIntosh, and Rome Beauty were five other important apple varieties developed at the turn of the 19th century (Figures 2.2, 2.3, and 2.4). King and Northern Spy were both from New York, which was beginning to lead the nation in apple growing. McIntosh came from Canada, Winesap was developed in New Jersey, and Rome Beauty was one of the first important varieties to be developed in Ohio. These varieties gained popularity during the 19th century and were still well regarded at the turn of the 20th century. McIntosh, a very hardy and thick-skinned apple, was well adapted to northern latitudes along with Northern Spy and King, whereas Winesap was a heat-loving apple rather like Ben Davis, and Rome Beauty was particularly good at higher elevations. Of these varieties, Northern Spy, McIntosh, and Rome Beauty would retain their perceived value for the longest duration. Northern Spy was used to create disease-resistant (Malling Merton) rootstocks in England in the 20th century, and McIntosh and Rome Beauty were two of very few 19th-century commercial apple varieties to remain economically important into the 21st century.

Role of Johnny Appleseed

The large number of American apple varieties emerging at the start of the 1800s was selected from a vast gene pool that had been seeded and spread throughout the early United States in seedling orchards. The legendary but real figure of Johnny Appleseed actually contributed to the vast dissemination of apple genes in this period, through his trade selling apple seedlings. Johnny Appleseed, whose real name was John Chapman, left his home state of Massachusetts in 1797 at the age of 23, to ply the rivers of the Northwest Territory in a dugout canoe. Using Allegheny County in western Pennsylvania as a base, Chapman paddled throughout Ohio and parts of Illinois and Indiana, sowing apple seeds that he derived from the pomace (mash) of cider mills.

Chapman cleared land areas beside rivers to establish nurseries of apple trees sown from seed, and later sold two-year-old apple seedlings to settlers on the frontier. He established numerous nurseries up and down the Ohio Valley, and attempted to locate his nurseries ahead of new settlement so that trees would be ready when settlers arrived. Noted by settlers as a rather unusual individual, he lived a shoeless, spartan lifestyle with a burlap shirt and tin pot hat, preferring to sleep out in the rough and commune with nature than live inside a dwelling. However, many of the legends of Chapman's colorful personality were probably fueled by an article in an 1871 issue of Harpers New Monthly magazine, "Johnny Appleseed: A Pioneer Hero," by W.D. Haley. Chapman was a self-ordained missionary of the Church of New Jerusalem, a Christian church based on the interpretation of the biblical writings of Emanuel Swedenborg, the Swedish scientist and theologian. Chapman and his fellow Swedenborgians believed that no

Figure 2.4: Watercolor paintings of Northern Spy (left), McIntosh (middle), and Rome Beauty (right) apple varieties (from Beach, 1905 courtesy of New York State Experiment Station).

separation exists between the natural world and the divine. In the words of writer Michael Pollan, Chapman had the "flinty toughness of Daniel Boone with the gentleness of a Hindu" (Pollan 01: 33).

When Chapman died in Fort Wayne, Indiana in 1845, his seedling apple tree nurseries were spread around four states on 1,200 acres of land that he had purchased from his income selling apple seedlings. Perhaps inadvertently, or by intention, he had managed to accelerate the mixing and spreading of millions of apple genes from the east throughout the former Northwest Territory. The apple varieties Black Annette, Franklin, Ohio Nonpareil, Western Beauty, and Ingram are believed to have been derived from the seedlings spread by John Chapman (Burford 07: n.p.).

Varieties Added to Farm Orchards

As wealthy landowners and common farmers began to recognize the benefits of American varieties in their adaptability to local growing conditions, the growing of true varieties became more accepted, and farmers began to learn how to graft their own trees. Scion wood of choice varieties was shared between farmers rather than bought from nurseries, and farmers taught their sons and extended family members how to graft a shoot to a young seedling tree (Figure 2.5). As the practice of grafting spread during the early 19th century, farm orchards of seedling trees were gradually replaced with grafted trees of true varieties, and new orchards were planted out with grafted trees rather than from seed. Coincidentally, the number of American varieties of apple and peach exploded into the thousands, so that single townships or counties had their own particularly well adapted variety that local farmers grew exclusively with pride.

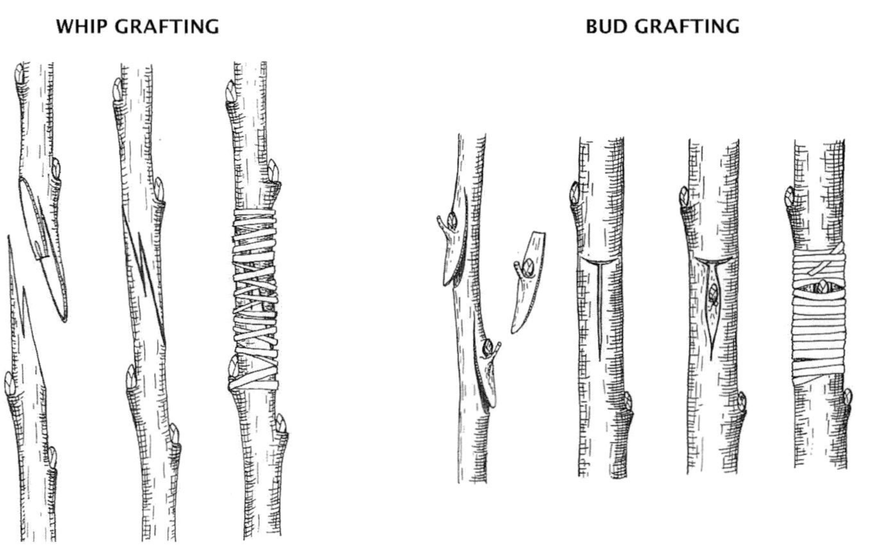

Figure 2.5: Diagram showing the steps involved in whip grafting versus bud grafting. The two grafting methods differ in the type and quantity of scionwood grafted to the rootstock (from Lowther, 1914, courtesy of the Encyclopedia of Horticulture Corporation).

In the state of New York, where development of apple varieties was matched only by the state of Massachusetts, the varieties grown were predominantly of local origin. For example, Northern Spy and McIntosh were grown in northern New York, King of Tompkins County was grown in central New York, Newtown Pippin was grown on Long Island, and Esopus Spitzenburg was grown in the Hudson Valley (Beach 05: I, 38). Peach varieties were still uncommon outside of fruit gardens from Pennsylvania to the South, though by the early 19th century some excellent American varieties had been developed. Perhaps the finest was the oldest, Heath Cling, though the others, Oldmixon Free, Cling, Lemon Cling, Red Rareripe, and Morris' White were regarded as good as any European peach variety.

Lack of American Pear Varieties

Before the turn of the 19th century, pear growing was limited to the northern states, and was localized in areas of French or Dutch influence, such as the Great Lakes regions of New York, Pennsylvania, Ohio, Michigan and Indiana, the Hudson Valley and Long Island areas of New York, and the Mississippi Valley areas of Illinois and Missouri. Pears did not thrive from Virginia to the South due to fire blight, though English settlers attempted to grow their familiar Bergamot and Warden (Pound) varieties with some success. Where pear trees could be grown well, they were in limited numbers, as pears grown from seed were highly variable in size, shape, color and flavor, and were often hard-fleshed and astringent in taste. Their uses were limited to cooking and to making pear cider or perry. The lack of popularity of the pear, and its greater genetic stability than the apple, meant the development of American pear varieties lagged far behind the development of the apple.

Rise in Pear Popularity

The lack of popularity of the pear changed during the first half of the 1800s when a pear-breeding craze in Belgium spawned a period of American enthusiasm that lasted from 1820 until 1860. During this period, a flood of new varieties and the development of the first American varieties led to the unparalleled popularity of the pear, before and since. Pear orchards were heavily planted throughout New York, New England, and the upper Midwest, and the fruits competed closely with apple in popularity. The dramatic increase in popularity of the pear in the early 1800s led Andrew Jackson Downing Jr., a New York gentleman farmer, landscape gardener, architect and horticultural writer, to describe the pear in 1847 as "…the favorite fruit of modern times and modern cultivators" (Downing 47: 316).

Pear breeding in Belgium was actually begun 70 years earlier, in 1730, by a priest named Nicolas Hardenpont in the town of Mons. Hardenpont was impressed by a 1694 text by Camerarius, which described the concept of sexual reproduction in plants for the first time. Hardenpont was inspired

Figure 2.6: Watercolor painting of Beurre du Bosc (Bosc) pear variety (from the USDA 19th and early 20th-century watercolor collection, courtesy of the USDA).

to sow and cultivate hundreds of seedling pears from French varieties, from which he selected and introduced 12 new varieties, including Passe Colmar. Hardenpont's passion for pear breeding was adopted by his fellow countrymen in the 19th century, when several thousand new varieties were created.

Between 1800 and 1850, Belgium became a country of pear breeders. Belgian priests, physicians, scientists, apothecaries, attorneys, tradesman, and gentlemen of leisure launched themselves into pear breeding, and were awarded medals for the creation of superior new varieties. Among the most remarkable breeders was Van Mons, who grew more than 80,000 pear trees from seed at his "Nursery of Fidelity" in Louvain (Beach 05: I, 40). Van Mons selected more than 400 new varieties from his nursery, and through his efforts and that of his contemporaries, the qualities of pear were transformed. The 19th century Belgian breeders improved the crisp or breaking flesh of existing pear varieties and created the soft, buttery pears, which were named for butter, or Beurre, such as Beurre d'Anjou, Beurre Bosc, and Beurre Gris (Figure 2.6).

First American Pear Variety

William Coxe was one of the first Americans to import Belgian pear varieties into the United States, and to begin their hybridization to create American varieties. Coxe was a gentleman farmer with land in Burlington, New Jersey, where he developed an experimental orchard and tested Belgian pears and European apples. In 1817 he published *View of the Cultivation of Fruit Trees*, an early American horticultural work of great importance, in which he described more than 65 Belgian pears that he had grown, and "...one hundred of the most estimable apple varieties in this country..." including Esopus Spitzenburg, Newtown Pippin, and Winesap. The most important aspect of his book, however, was his description and promotion of one of the first pear varieties developed in the United States, the Seckel (Figure 2.7).

Figure 2.7: Watercolor painting of Seckel pear variety (from the USDA 19th- and early 20th-century watercolor collection, courtesy of the USDA).

The development of the Seckel pear along the Delaware River south of Philadelphia had occurred just a few years before Coxe wrote his book. As *View* was published, Seckel was available for sale in Philadelphia for the first time. Coxe did much to influence the popularity of Seckel with American pear orchardists. He described it as "...the finest pear in this or any other country..." Seckel is an early fall-ripening, medium sized pear, with a yellow-green, red-blushed skin. Its concentrated, spicy and honeyed flavor soon placed the fruit in great demand throughout the Northeast, and it was one of the first American fruits to be held as an emblem of nationalistic pride, along with the Newtown Pippin apple variety.

American Fruit Breeding

For the next several decades, specialist fruit growers—pomologists—monopolized the work of fruit breeding and variety selection on American soil. Centered in the Northeast, pomologists were generally gentlemen farmers with other professions, such as medicine or law, who had a passion for fruits and were well-read in European horticultural literature. Rather like William Coxe, they used their wealth and leisure time to develop experimental orchards upon their land, and create new fruit varieties by growing hundreds and sometimes thousands of trees from seed and selecting trees that bore exceptional fruits. These men had access to European fruit varieties through nurseries in large cities, or through direct contacts in Europe. European varieties were often the raw materials for their fruit breeding work. The pomologists were familiar with each other's work through various societies and published treatises, and were often associated with the London Horticultural Society, the English clearinghouse for European fruit varieties.

Early American Pomologists

Among the important figures in pomology in the first half of the 1800s was William Kenrick of Newton, Massachusetts, Robert Manning of Salem, Massachusetts, David Hosack of New York, Marshall Wilder of Boston, and William Hamilton of Philadelphia. These men imported European varieties of all fruit species, though mostly pear, and described the best varieties, selected new American varieties, and distributed them to nurserymen. David Hosack, a New York physician, introduced Seckel to the London Horticultural Society, whose collections included hundreds of new Belgian pear varieties. Seckel was received with great praise by English pomologists, who described it as one of the best fall-ripening pears. Robert Manning imported all of Van Mons' pear varieties into America, and described these and others in his 1838 *Book of Fruits*. When Manning died in 1842, he left behind more than 2,000 varieties of fruits in his pomological garden in Salem.

First Native Fruit Cultivar

As interest in fruit breeding spread into newly settled areas, the first variety or cultivar of an American native plant was developed. In 1814, the native Chickasaw plum, *Prunus angustifolia*, was hybridized on an estate in Knox County, Kentucky, to create the variety Miner. The Miner plum had larger fruits than the native species, with more tender skin, and softer, sweeter flesh. The new variety stimulated the growing of plums in the southern states, where European plums were less well adapted to the hot summers. The development of a variety from a native species heralded great promise for the future of American fruit growing, and inspired the development of a large number of native plum varieties in the 19th century.

First Commercial Orchards

The first commercial orchards in the United States were planted in the Hudson Valley and on Long Island, New York. The advent of the steam engine in the late 18th century and its application to power steamboats in the early 19th century, led to the first rapid means of transporting perishable goods in America as well as the development of agricultural commerce along the eastern seaboard. Among the first settlers of these areas were French Huguenots and Dutchmen, who brought horticultural skills from their countries where the French system of intensive horticulture was flourishing in market gardens by the 18th century. The ethnic horticultural skills of the settlers and their proximity to the largest port in the country meant that eastern New York was the first land area to be planted with orchards of cultivated varieties, and the quality of their fruits was superior to most others at the time.

Access to transportation and a burgeoning destination for market produce, along with growing conditions suitable for many orchard fruits, stimulated establishment of commercial orchards along the Hudson River and the northern coast of Long Island during the 1820s. These orchards were started from grafted trees rather than seed, and were laid out with a regular arrangement of trees. As the land area of the common farmer was limited before the Homestead Act of 1862, and the mature size of many varieties was unknown at the time, apple and pear trees were planted densely at 20-30 feet apart in rows 20-30 feet apart. Newtown Pippin and Esopus Spitzenburg were among the first commercial apples to be sold in New York City, where they were received as a novelty.

As the demand for dessert apples grew and the profitability of commercial orcharding became known, commercial orchards spread into upstate New York and along the Niagara River Valley, where fruits could be transported via the Erie Canal to the Hudson River and New York City. The building of canals and railroads gave farmers in interior locations throughout the Northeast and mid-Atlantic states access to small, local markets, where they could sell their produce for profitable returns. During the first half of the 1800s, many farmers switched from vegetable crops or farm orchards to commercial orchards, relying on fruit varieties, steamboats, trains, and new wagon roads to make a living from apples, pears, and peaches.

Advent of Agricultural Journalism

The adoption of commercial fruit growing by common farmers rather than educated pomologists created a market for agricultural journalism that would convey horticultural news and information in a populist style. Beginning in the 1820s, agricultural journalism helped to educate a generation of farmers in the skills of fruit growing, through newspapers, newsletters and magazines such as *The Cultivator* and *American Farmer*. Nursery catalogs became valuable sources of information for orchardists, and the Prince Nursery catalogs printed between 1815 and 1859 were among the most important horticultural publications of the time. The demand for cultivated varieties fueled the development of nurseries outside of the major cities, and as the United States grew with the Louisiana Purchase of 1803, nurseries and orchards emerged in the central Midwest.

Advent of Stark Brothers Nursery

It was in this climate of enterprise and expansion that the Stark Brothers Nursery was founded in Louisiana, Missouri, in 1816. Stark Brothers began by propagating and selling eastern and European fruit varieties that could be grown successfully in the Midwest. The Starks found great opportunity in the Midwestern climate, where a greater range of fruit species could be grown than in the north or south. Certain varieties of apples and pears could be grown along with peaches and plums, and this range provided a lucrative advantage to the young nursery. During the 19th century, the Starks expanded their scope by developing new varieties for the Midwest and selling all the important commercial varieties in the United States. In the early 20th century, Stark Brothers became one of the most significant nurseries in the history of fruit development when it patented the Delicious and Golden Delicious apple varieties.

Development of Export Trade

With the end of the War of 1812 came the opportunity for the development of an export trade with England. The first documented export statistics of the United States Treasury are from 1821, when 68,443 bushels (1,437 tons) of fresh apples were shipped to England (Beach 05: I, 64). An export market for dried apples developed too, though both fresh and dried transatlantic apple shipments did not become economically significant until after 1838, when Queen Victoria repealed the tax on imported apples. The export pioneer Newtown Pippin appeared at the center of the story once again. In 1838, Andrew Stevenson, a Virginian and the American Minister to the Court of St. James, presented Queen Victoria with two barrels of Virginia-grown Newtown Pippins. Legend has it that the Queen was so impressed with the apples that she abolished import duties on American apples, and the export trade in fruits began in full force (Hatch 98: 72).

Pacific Northwest Introductions

By the 1820s, farm orchards were planted from the north to the south, east of the Mississippi River, and to a limited extent, in the newly acquired lands of the Louisiana Purchase, which included the lands between the Mississippi River and the Rocky Mountains. Spanish and Mexican lands in the Southwest had been planted with a mixture of Mediterranean and temperate fruits in missions and ranchos, and the northwest lands owned by Great Britain were being explored for animal furs and other natural bounties. During the 1820s the British Hudson's Bay Company established an agricultural outpost near the mouth of the Columbia River in the Oregon Territory (in present-day Washington State). Over the next 20 years the outpost grew to encompass many thousands of acres and served as the main food supply for company members in the Northwest.

In 1825 the first apple tree was planted at the outpost known as Fort Vancouver, originating from a seed brought from England. A decade later the agricultural and horticultural activities of Fort Vancouver were flourishing when some of the first Euro-Americans to cross the Rocky Mountains arrived in search of hospitality. Narcissa Whitman, the wife of the missionary Dr. Marcus Whitman, recorded the following observations in 1836, when she stayed at the Fort with fellow missionaries from the east, Reverend Henry Spalding and his wife Eliza:

> *Here we find fruit of every description, apples, peaches, grapes, pears, plums and fig trees in abundance; also cucumbers, melons, beans, peas, beets, cabbages, tomatoes and every kind of vegetable, too numerous to be mentioned. Every part is very neat and tastefully arranged, with fine walks lined on each side by strawberry vines. At the opposite end of the garden is a good summer house covered with grape vines...*

Hudson's Bay Company Influence

Before the United States acquired the land between the 45th and 49th parallels from Great Britain in the 1846 Oregon Treaty, the Hudson's Bay Company established several outposts in the Oregon Territory, where it planted fruit trees and traded fruits and seeds with American Indians (Figure 2.8). The first farm orchard planted by an American Indian in the Pacific Northwest is believed to have been at the mouth of Alpowa Creek in Asotin County in 1837 (in present-day Washington). The apple orchard was sown by a Nez Perce Indian, Red Wolf, with seed obtained from the Hudson's Bay Company.

Oregon Trail Migration

The Hudson's Bay Company's influence on fruit dispersal in the Pacific Northwest ended in 1846 when the Oregon Territory was opened up to settlement by citizens of the United States, who brought fruits from east of the Rocky Mountains. One Homeric saga in the westward migration of orchard fruits featured two nurserymen who met in Iowa in 1846, named Henderson Luelling and William Meek. Determined to pursue a lucrative business opportunity, Luelling and Meek individually hauled many hundreds of young grafted fruit trees of choice varieties along the Oregon Trail in 1847. The men reconnoitered in Oregon City in the Willamette Valley, where they combined their stock and established the first fruit tree nursery in the present-day Pacific Northwest. The Luelling and Meek Nursery, located just outside the territory's new capital city, supplied all of the first orchards of Donation Land claimants. The Donation Land Act of 1850 provided for a grant of land to settlers in Oregon. The law prescribed residence on and cultivation of the land in order to secure a patent and gave 320 acres to each single man and 640 acres to a married couple.

The remarkable story of Luelling's journey was transcribed by the Oregon State Historical Society from details provided by Luelling's son Alfred, and was published in *the Proceedings and Papers of the Quarter Centennial Celebration of the Oregon State Historical Society* in 1910:

> *Early in April, 1846 a critical selection was made of between 800 and 900 of the best trees, one year old from the graft, and they were planted in two wagon boxes in a foot of earth in which was uniformly mixed a lot of pulverized charcoal. A light framework was built upon the top of each box in order to protect the young and tender limbs from injury by stock. Each wagon was drawn by four yoke of oxen, Mr. Luelling driving the first team, and his son, Alfred, a lad of 16, the second team. The trees were watered on the plains as the opportunity offered, and the precious load arrived at Portland—then a hamlet containing not to exceed 20 rude log cabins, without any loss worth mentioning. . .Thus was the first nursery established on the Pacific coast. That load of trees contained health, wealth and comfort for the old pioneers of Oregon.*

Figure 2.8: Historic photograph of a Klickitat Indian woman with a seedling apple tree obtained from the Hudson's Bay Company in Washington Territory in the early 1800s (from Lowther, 1914, courtesy of the Encyclopedia of Horticulture Corporation).

Luelling and Meek were careful to select varieties that would spread the period of fruit harvest from summer through winter. Among the summer apple varieties Luelling hauled to Oregon were Summer Pearmain, Golden Sweet, and Red Astrachan (Figure 2.9). The fall apples included Gravenstein, Westfield Seek-No-Further, and King of Tompkins County (Figures 2.10 and 2.11). The winter apples had the broadest selection, with the varieties Yellow Belleflower, Baldwin, Newtown Pippin, Esopus Spitzenburg, Lady Apple, Northern Spy, and Winesap among the precious cargo (Figures 2.12, 2.13, and 2.14). Luelling also selected summer, fall, and winter ripening pears. Bartlett was among the summer pears, and Seckel and Flemish Beauty were for the fall (Figure 2.15). Winter Nelis, one of Luelling's winter ripening pear varieties, would become economically important in the Northwest in the early 20th century, along with Bartlett (Figure 2.16).

Figure 2.9: Watercolor painting of Red Astrachan, a summer-ripening apple variety (from Beach, 1905 courtesy of New York State Experiment Station).

Figure 2.10: *(right)* Watercolor painting of Gravenstein, a fall-ripening apple variety. Note the variety typically has more green and speckling (from Beach, 1905 courtesy of New York State Experiment Station).

Figure 2.11: *(below)* Watercolor painting of Westfield Seek-No-Further, a fall-ripening apple variety (from Beach, 1905 courtesy of New York State Experiment Station).

Figure 2.12: Watercolor painting of Yellow Belleflower (left) and Baldwin (right), two winter-ripening apple varieties (from Beach, 1905 courtesy of New York State Experiment Station).

Figure 2.13: Watercolor painting of Newtown Pippin (left) and Esopus Spitzenburg (right), two winter-ripening apple varieties (from Beach, 1905 courtesy of New York State Experiment Station).

Figure 2.14: Watercolor painting of Lady, a winter-ripening, tiny apple variety (from Beach, 1905 courtesy of New York State Experiment Station).

Figure 2.15: *(lower left)* Watercolor painting of Flemish Beauty pear variety (from the USDA 19th and early 20th-century watercolor collection, courtesy of the USDA).

Figure 2.16: *(lower right)* Watercolor painting of Winter Nelis pear variety (from the USDA 19th and early 20th-century watercolor collection, courtesy of the USDA).

Important 1800s Apple Varieties

In 1830 two more commercially important apple varieties were developed that would retain their significance until the 20th century: Jonathan and York Imperial (Figures 2.17 and 2.18). Jonathan was developed in the Hudson Valley of New York from a seed of Esopus Spitzenburg. A yellow-red apple with a mild acid flavor, Jonathan is a fall-ripening variety that was planted throughout the Great Lakes region of the upper Midwest and Pennsylvania, where it became one of the most important September apples. Curiously, Jonathan managed to survive the rigorous commercial selection pressures of the 20th century, and is one of the only commercial non-winter apple varieties still grown in the United States at the beginning of the 21st century. York Imperial, originally named Johnson's Fine Winter, is a distinctive white-crimson striped apple. It was developed in Pennsylvania and became an important commercial variety in southern orchards, from Virginia west to Arkansas. York Imperial remained popular until the mid-1900s when many of the last surviving varieties from the 1800s were dropped from commercial production.

Figure 2.17: *(left)* Watercolor painting of Jonathan, a fall-ripening apple variety (from Beach, 1905 courtesy of New York State Experiment Station).

Figure 2.18: *(left)* Watercolor painting of York Imperial, a winter-ripening apple variety (from Beach, 1905 courtesy of New York State Experiment Station).

American Varieties Begin to Dominate

The year 1830 is a landmark in the history of fruit development in the United States, as in that year, the Prince Nursery catalog offered as many American varieties of apple for sale as European varieties. The rapid development of American apple varieties, their excellent quality and better adaptability to local growing conditions, all served to out-compete the European apple varieties, which were starting to diminish in popularity. American peach varieties were also gaining an excellent reputation throughout the United States and abroad, and only pear, cherry, and plum varieties would remain relatively undeveloped, with European or Asian varieties being dominant. In 1841, the Prince Nursery catalog offered 272 varieties of apple, 420 varieties of pear, 109 varieties of cherry, and 156 varieties of plum. By 1847 the balance between the availability of European and American peach varieties had also shifted, with more American peach varieties for sale than European, for the first time. The 1847 Prince Nursery catalog offered 76 varieties of peach, 48 of which were American varieties and only 28 were European.

Influence of Andrew Jackson Downing

By the mid-1800s, American fruit varieties were a significant source of national pride. With signs of strain on the Union becoming evident, symbols of national identity were valued as beacons of hope for the future. During the 1800s the apple became a symbol of American democracy, and horticultural writers took advantage of opportunities to praise the superior qualities of American apples and other American fruit varieties. One such booster was Andrew Jackson Downing Jr., whose book *The Fruits and Fruit Trees of America*, first published in 1847, would become the most widely read horticultural text in 19th century America, and the standard authority in fruit growing for the next 50 years.

Downing grew up in the Hudson Valley, where he became fascinated by orchards and studied fruit varieties in depth. His remarkable breadth of skills lent themselves to landscape design, architecture, journalism, and farming. Downing took a scientific approach to farming, and carefully documented his endeavors. He grew many different orchard fruits, and closely recorded their characteristics. These records and observations were compiled into *Fruits*, the most comprehensive account of the best fruit varieties grown in America at the time.

Soon after he wrote *Fruits*, Downing's life ended prematurely in a tragic steamboat accident. Despite this, his influence continued posthumously through 20 revised editions of the book between 1847 and 1900. Downing's brother Charles, another respected pomologist, contributed to several revised editions of *Fruits*. Downing's influence on fruit growing was considerable. He engendered confidence in American varieties, influenced the types of varieties grown, and taught Americans how to grow fruit trees. Of the Seckel pear, he noted "...the high flavor of the Seckel pear, an American variety, as yet unsurpassed in this respect, by any European sort, proves the natural congeniality of the climate of the northern states to this fruit" (Downing 47: 317). The Newtown Pippin, he said, "...is pretty generally admitted to be the finest apple in the world" (Downing 47: 56).

In 1847 Downing recommended several hundred varieties of fruit, including 159 varieties of apple, mostly American, with 24 summer-ripening varieties, 45 autumn varieties, and 90 winter varieties. The broad range of excellent apple varieties meant that where a number of varieties were grown, fresh apples could be picked from June through November, and the last ripening varieties could be stored from winter through spring, almost until the next June crop appeared on the trees. Downing also recommended 241 varieties of pear, the majority of which were European and fall ripening, rather than summer or winter. He also recommended 76 varieties of peach, the majority of which were American, and many cherry and plum varieties that were mostly European.

Downing's Influence on Cultivation

In addition to promoting particular varieties of fruit, Downing had a great influence on horticultural methods and the appearance of American orchards during the latter half of the 19th century. He advocated generous spacing for fruit trees, good soil preparation and sanitation practices, neatness and orderliness, but differed from his pomologist counterparts in his reluctance to prune orchard trees. His reluctance was based on a perception of a firm distinction between orchards and fruit gardens. Orchards, he thought, should be plantations of large, unpruned, well spaced trees, grown to take up their full size and shape, whereas fruit gardens were smaller enclosures of closely spaced, highly pruned and sometimes dwarfed, fruit trees (Figure 2.19). Downing believed a lack of pruning helped to promote the longevity of the orchard. Longevity was a concept of great appeal to farmers of the time. Downing encouraged farmers to plant an orchard as a family legacy, an investment they could expect to pass on to their sons and be profitable for 50 to 75 years.

Origins of Dwarf Trees

While dwarf apple and pear trees were available from finer nurseries in the mid-1800s, Downing did not express a great deal of interest in them. Nurserymen created dwarf apple trees by grafting the stems of choice apple varieties onto the roots of one of two imported apple varieties, Doucin or Paradise, also known as French Paradise and English Paradise respectively. Doucin and Paradise were available to nurserymen throughout the 19th century, though their importance was limited until the 1900s when English

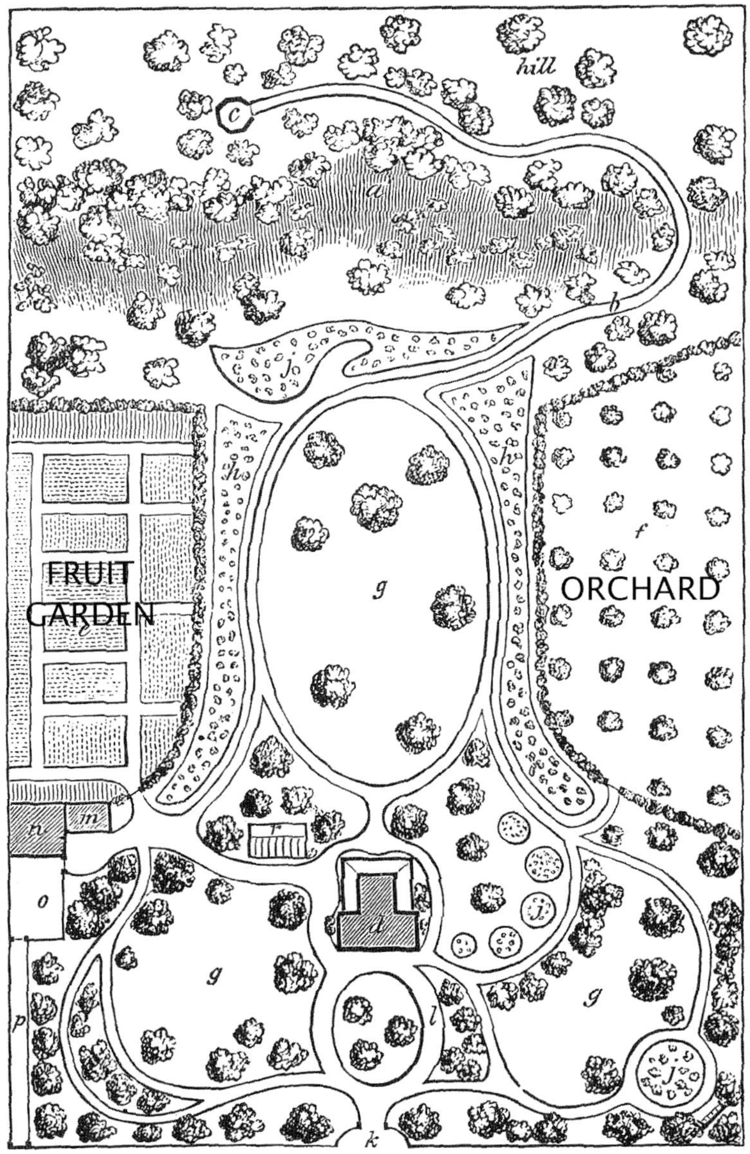

Figure 2.19: Drawing of a mid-1800s gentleman's farm, Bracket Farm, designed by Andrew Jackson Downing Jr., showing the fruit garden distinct from the orchard (from Downing, 1852, courtesy of J. Wiley Co.).

horticulturists used them to develop a system of commercial dwarfing rootstocks, thereby transforming the appearance of apple orchards.

Dwarf pear trees could also be purchased in the mid-1800s, and were created by grafting a choice pear variety onto the roots of the common quince. Like Doucin and Paradise, the quince rootstock retarded aerial growth and produced a tree less than half the full or standard size. Dwarf apple and pear trees were uncommon, however, and were, in Downing's words "the domain of the gardener, rather than the orchardist." Dwarf fruit trees were more expensive than standards, produced smaller sized fruit, and were short-lived. Downing's interest was clearly in the orchard rather than the fruit garden, and with standard rather than dwarf trees.

He perceived the audience for his book to be the American farmer rather than the gentleman pomologist, as he declared his thoughts on pruning: "Every fruit tree grown in the open orchard or garden as a common standard should be allowed to take its natural form, the whole efforts of the pruner going no further than to take out all weak and crowded branches" (Downing 47: 56).

Orchard Characteristics of Period

Downing recommended a 30-foot grid spacing for apple and pear trees, and a 16- to 20-foot grid for peaches, cherries, and plums. This was wider spacing than typically used at the time. He promoted the grafting or budding of trees very close to the ground, so that the scar of the graft union would appear only 3-4 inches high on the trunk. He also believed in "heading high" the tree canopy, so that the branches were borne on a relatively tall trunk (Figures 2.20 and 2.21). A higher canopy kept the precious fruits out of reach of grazing cattle and swine, but such lack of access also presented challenges for human harvesters. This apparently was not a concern for Downing or his contemporaries, as farm labor was plentiful.

Throughout the latter 19th century, a companion industry of harvesters' accessories developed to supply equipment for picking fruit from very tall trees, commonly 20-30 feet above the ground. These included very tall pickers' tripod tree ladders, and various styles of leather aprons for pickers, with built-in containers for holding fruit. The aprons' false bottoms eased the transfer of picked fruit into the waiting barrel, after the picker had negotiated the ladder descent, with full apron intact.

Figure 2.20: Drawing showing 19th-century equipment for transplanting a fruit tree, depicting a typical tree form with a characteristically tall trunk (from Farmer's and Housekeeper's Encyclopedia, 1888, courtesy of F. M. Lupton).

Figure 2.21: Drawing of a 19th-century "model apple tree," with the characteristics of a tall trunk, a very large canopy, and an unpruned form—seen in the crowded scaffold branches (from Bailey, 1888, courtesy of Orange Judd Company).

Biennial Bearing

Downing also advocated preventing older fruit trees' natural tendency to bear fruit biennially rather than annually by thinning out the developing fruit. This technique prevented the production of a large crop one year and a small crop in the following year. The tendency of older fruit trees to become "biennially bearing" is not a product of resource depletion, but is a hormonal response as a result of seed production in the developing fruit. The need to control biennial bearing for more consistent cropping and a desire to dwarf trees for easier access, were two issues that would contribute to the transformation of orchards in the 1900s.

Ubiquity of Fruit Growing

Andrew Jackson Downing Jr. wrote *Fruits* at a time when more farmers were taking up commercial orcharding as a vocation than at any other time in American history, and when other farmers were planting at least a handful of fruit trees for subsistence and/or curiosity. By 1847, in Downing's words, "...every farmer is an amateur fruit grower and connoisseur." His book reached farmers throughout the east, the rapidly growing Midwest, and as far west as Utah, where Mormons were beginning to establish farming settlements with orchards of true varieties watered by irrigation systems. For the next 50 years, as the many editions of *Fruits* were published, the typical apple and pear orchard would have a 30-foot grid spacing, with very large, almost entirely unpruned trees, bearing high canopies on three- to six-foot-tall trunks (Figures 2.22 and 2.23). Following Downing's recommendations, orchards were planted with single varieties in rows for ease of harvest and a neat and orderly appearance, rather than to facilitate cross-pollination, which was not understood until the 1920s.

Figure 2.22: Photograph of a 19th-century Baldwin apple tree with a characteristically tall trunk, Adams National Historical Park, MA (Heidi Cope, Horticulturist, for scale) (S. Dolan, 2001).

Pest Control in Period

Until the advent of pesticides in the 1880s, pest control was performed by grazing livestock or by hand, though few farmers adhered to Downing's laborious recommendations for sanitary practices. Pests and diseases were not rampant in Downing's time, but became widespread during the latter 19th century, even reaching epidemic proportions. Pests and diseases were spread through the expansion of agricultural settlement, the destruction of native host plants (including forests), the importation of plants and fruits, and the mass planting of orchards throughout every part of the country. The epidemics ushered in a new era of scientific horticulture in the 1880s. Until this time, farmers practiced horticulture without scientific knowledge of disease theory and biochemistry, and Downing's relatively innocuous recommendations for pest control were typical of those available based on personal hygiene.

Figure 2.23: Photograph of a 19th-century McIntosh apple tree with a characteristically tall trunk and large size, Oxon Hill Farm, DC (Jim Rosenstock, National Capital Parks East Ranger, for scale) (S. Dolan, 2001).

Downing advocated for the manual scraping of twigs and trunks to remove insects, and the washing of trees with soapsuds to remove pest residues. He suggested a sponge dipped in ammonia and borne aloft on a long stick to douse nests of bagworms and other pests, and recommended a canvass bandage painted with tar for the tree trunk, to resist the migration of soil-borne, crawling insects. Bonfires lit in the orchard at night could destroy codling moth (which was becoming a recognized problem), and a mixture of manure and urine, applied to the soil as a tonic, was a stock remedy for most ills.

Downing struggled to recommend treatments for problems not apparently caused by insect pests. Without knowledge of microorganisms, Downing could offer no apparent causal agent for fire blight, a bacterial disease that was becoming a problem with pears in the Northeast. He could only suggest the removal of blighted limbs and the subsequent painting of pruning wounds with liquid shellac or white lead. A severely diseased tree, he said, should be felled and then burned. Although he could do little to stem the tide of infestations that would overwhelm many commercial orchards by the 1880s, Downing's profound influence on American horticulture was to promote the planting of a multitude of fruit varieties in orchards of a type that would eventually become an archaic idiom.

Advent of American Pomological Congress

Before his death in 1849, Downing left behind another legacy. In 1848 Downing and his brother, Charles, helped to found the American Pomological Congress in Buffalo, NY. The Congress was composed of a number of prominent pomologists, who made their charter the identification and description of fruit varieties, including the identification of synonyms and the testing and recommendation of the best commercial varieties. In 1852, the American Pomological Congress was renamed the American Pomological Society, and published its first report, ***Fruits Worthy of Cultivation***. The report contained a list of the 32 most highly recommended apple varieties of the day. The list identified the three best early-fruiting apple varieties as Summer Pearmain, Early Harvest, and Early Strawberry; also New England's favorite, Baldwin, and four very cold-hardy varieties, Fameuse, Gravenstein, Hubbardston, and Red Astrachan (Figures 2.24, 2.25, 2.26 and 2.27). The old varieties Roxbury Russet, Rhode Island Greening, and Westfield-Seek-No-Further also made the list, along with Lady Apple, Summer Rose, Swaar, and Winesap. For specific locations, the popular varieties Esopus Spitzenburg, Newtown Pippin, Northern Spy, and Yellow Belleflower were recommended.

Figure 2.24: *(left)* Watercolor painting of Early Harvest, a summer-ripening apple variety recommended as one of the best apples in 1852 by the American Pomological Congress (from Beach, 1905 courtesy of New York State Experiment Station).

Figure 2.25: *(left)* Watercolor painting of Early Strawberry, a summer-ripening apple variety recommended as one of the best apples in 1852 by the American Pomological Congress (from Beach, 1905 courtesy of New York State Experiment Station).

Figure 2.26: *(left)* Watercolor painting of Fameuse, a winter-ripening apple variety recommended as one of the best apples in 1852 by the American Pomological Congress (from Beach, 1905 courtesy of New York State Experiment Station).

Figure 2.27: *(left)* Watercolor painting of Hubbardston, a winter-ripening apple variety recommended as one of the best apples in 1852 by the American Pomological Congress (from Beach, 1905 courtesy of New York State Experiment Station).

Taxonomy of Apple and Pear

Several members of the American Pomological Society had considerable influence over the popularity of the pear in the 19th century, most notably Wilder, Hamilton, Mannings Sr. and Jr., and the Downing brothers. During its heyday from 1825 to 1860, the pear was considered to be the fruit of connoisseur taste, and was generally regarded by the populace as a more precious commodity than the apple. Hundreds and possibly thousands of varieties became available, the majority being European varieties, though some were American, and almost all were originated from the European pear species, *Pyrus communis*. This pear species, like the species of the domestic apple, *Malus domestica*, is now considered by taxonomists to not be naturally occurring, but rather to be the prehistorically man-made hybrids of species that are naturally occurring, namely *Pyrus caucasia* and *P. nivalis* for the European pear, and possibly *Malus floribunda*, *M. zumi*, *M. sargentii*, and *M. sieboldii* as the wild parents of the domestic apple (Reisner 2000). Only one other pear species has economically important fruit—the Asian pear, or *Pyrus pyrifolia* [syn. *P. serotina*], also known as the Japanese, Asian, or even Chinese Sand pear.

Introduction of Asian Pear

The Asian pear is shaped more like the apple, with a yellow or russetted skin. It has crisp fruit with a granular texture, and a low-acid, watery-sweet taste. During the heyday of pear varieties derived from the European species, such as Bartlett, Beurre D'Anjou, Beurre Bosc, and Comice, the Asian pear was introduced to the United States via the Prince Nursery, through its connections with the Royal Horticultural Society of England. The Society had imported the Asian pear from Japan in 1820, and made it available to the Prince Nursery in 1840. Immediately American pear breeders set to work to create American hybrid varieties between the Asian and European pear. They aimed to create a hybrid that combined the most favorable characteristics of the two species, the succulence and melting flavor of the European pear, with the heat, drought and blight resistance of the Asian pear. A good American hybrid variety between the two species was thought to be potentially lucrative in the southern states, where varieties derived from the European pear failed in the hot summers.

Within six years, the first hybrid variety was on sale at the Prince Nursery, namely the Le Conte, for its originator Major John Le Conte of New York (Figure 2.28). Le Conte and other new hybrids were slow to gain popularity in the coming decades, however, as the fruits were often dry and gritty. The variety Kieffer was developed in 1873, superseding Le Conte, and causing the economic potential of an Asian hybrid pear to be seriously considered (Figure 2.29). By the early 1900s, Kieffer was acknowledged as the most suitable pear variety for the southern states, and was widely grown throughout the South, including Texas, and in the Midwest.

Euro-American Introductions to California

With the Mexican Cession of 1848 at the end of the Mexican American War, California, Arizona, Nevada, Utah, and parts of New Mexico and Colorado were admitted to the United States. Prospectors poured into the Sierra Nevada Mountains by the thousands, drawn by the California Gold Rush of 1849. Many empty handed placer miners soon discovered greater prospects in the potentially rich agricultural valleys and foothills of California, rather than mineral ore. Some of the first newcomers were fortunate to find the bones of a profitable orchard already laid out for them, in the abandoned orchards of Spanish missions, secularized less than two decades before. Some enterprising settlers like pioneer W.M. Stockton laid claim to Spanish mission grounds and attempted to rehabilitate the 75-year-old orchards. Stockton claimed the former San Gabriel Mission, and by 1853 had grafted the stumps of the old Spanish pear trees with his own stock varieties, establishing the first commercial pear orchard in California.

Figure 2.28: *(below, left)* Watercolor painting of Le Conte pear variety, the first American-bred European/Asian-hybrid variety (from the USDA 19th and early 20th-century watercolor collection, courtesy of the USDA).

Figure 2.29: *(below, right)* Watercolor painting of Kieffer pear variety, an American-bred European/Asian-hybrid variety (from the USDA 19th and early 20th-century watercolor collection, courtesy of the USDA).

West Coast Commercial Orchards

Within 20 years of the Gold Rush, California had a booming commercial orchard industry. California orchardists wasted no time in exploiting the completion of the transcontinental railroad in 1869. That year, the first commercial shipment of California apples and pears was made via rail to eastern markets. During the next two decades, the railroad penetrated far into the western United States and brought commercial orchards to the Wenatchee and Yakima Valleys of eastern Washington in the 1870s. The Euro-American settlers of eastern Washington had moved west with knowledge of fruit growing based on 50 years of commercial orcharding in the east. They immediately recognized the potential of the clear, sunny climate, hilly topography, and highly fertile soils for fruit production, and set about an industrial-scale transformation of the landscape into commercial orchards.

While much of the newly settled agricultural lands of the West were immediately developed as commercial orchards, small subsistence orchards were still being planted in the 1870s. The subsistence or farm orchard of few true varieties and scraggly seedling trees became a familiar signature of the western vernacular landscape after the passage of the Homestead Act in 1862. These orchards were often sited on infertile soils, full of recently cut forest stumps, with no apparent consideration to layout, spacing or means of irrigation. The curious state of these orchards was a manifestation of the Homestead Act, which inadvertently encouraged orchard planting in an arbitrary or speculative manner.

Influence of the Homestead Act

One of the provisions of the Homestead Act gave property title to claimants who could demonstrate their agricultural improvement of the land over a five-year period. A farm orchard, as a one-time planting effort, was a more economic form of cultivation in terms of money, labor and materials than annual planting of farm crops. After planting, occupation of the land to protect proprietary rights became unnecessary. The orchard could be abandoned for five years until a claimant received their Homestead Patent, when they would find it worthwhile to invest in the building of a dwelling. Often these neglected orchards were improved once the new owner occupied the property, with the addition of variety fruit trees. However, many "homestead orchards" were never reclaimed after pioneers abandoned their wait, and they were left to decline and scatter their seeds.

Squatters residing in the Yosemite Valley attempted to legitimize their possession of lands during the 1860s by planting orchards. As observed by C. Todd Kennedy, a California pomologist, homesteading actually led to the overplanting of the valley in fruit trees during this period. Irrational land use resulted in the excessive conversion of meadow and forest to orchard cultivation (Kennedy 91: 3). In 1991 Kennedy examined Yosemite Valley and found seven remaining orchards and more than several hundred fruit trees whose origins could be traced to the homesteading period. While many of the trees were seedlings, Kennedy also identified many variety trees, such as the apple trees found in the Curry Village Orchard (Figure 2.30). This remarkable orchard was planted in the early 1860s by James Chenowith Lamon, one of the first Euro-American settlers in Yosemite Valley who sold fruits and vegetables to early travelers and tourists.

End of Golden Age

The end of the "golden age of pomology" was marked by a loss of passion for the novelty of fruit growing for the sake of it. A period of excess in fashion, discourse, breeding, fascination, and mysticism was ended with orchard pest and disease infestations and their ensuing economic losses. Preoccupation and pride in the beauty and taste of fruits would be superseded in the 1880s with more pragmatic concerns for yield and economic viability. The end of one era was supplanted by the dawning of another, as scientific research entered the orchard and engendered a type of "industrial revolution" in fruit growing.

Figure 2.30: Photograph of nearly 150-year-old, mixed variety apple trees showing characteristically unpruned scaffold form, tall trunks, and large size, Curry Village Orchard at Yosemite National Park, CA (S. Dolan, 2006).

Summary

As the 1870s drew to a close, commercial orchards and farm orchards had been planted throughout the 48 contiguous states. Many thousands of true varieties of fruits were planted, with the greatest numbers of apple and peach being American varieties, and the majority of pear, cherry, and plum being European. Many farmers in the east had converted their fields from cereal grains and vegetables to orchards, and by this time most orchard fruits were grown for human consumption as fresh fruit. Other commodities were replacing the use of seedling fruits for cider and livestock feed.

By the end of the 1801-1880 period in the West, many newly claimed lands were immediately developed as commercial orchards. Orchards were laid out with relatively wide spacing between trees, typically 30 feet square for apple and pear, and 16-20 feet square for peach, plum, and cherry. The trees were grafted close to the ground and were allowed to develop tall trunks. A typical orchard tree had a large, unpruned canopy, and without fruit thinning by the orchardist, would bear a good crop only every other year.

Dwarf apple and pear trees were available through the nursery trade, which had spread to every larger city in the country. However, dwarf trees were generally found only in fruit gardens, which remained distinct from farm orchards or commercial orchards. Seedling fruit trees were still being sown in the newly settled West, though even the most remote farm orchards were typically laid out with a handful of true varieties in addition to seedling trees, to provide some good fresh fruit.

Ornamental plants such as rose, lilac, mock orange, and spirea were not available from plant nurseries until the 1860s. Until this time, American horticulture was defined by the growing of fruits, and fruit trees were grown for their ornamental qualities as well as for their bounty.

The threat from pests and diseases was formally realized as insects and pathogens traveled west with migrating people and new agricultural settlement. Pests that had been noted in the late 1700s in New York State had arrived in the Pacific Northwest by the 1880s. Fire blight had decimated pear orchards in the Northeast, peach yellows had infected peach orchards of the east and south, and codling moth and apple scab had blighted the apple orchards of Missouri, Kansas, Arkansas, Iowa, Mississippi, and Texas. Pest control methods were rudimentary, and at best, were based on notions of sanitary hygiene rather than science. Few alternatives for effective pest management were available.

Chapter 3
Orchard Specialization and Industrialization, 1881-1945

Commercialization, scientific, and technological development, and regionalism

Modern Era

The modern era of orchard fruit growing from 1881 to 1945 is characterized by increased technological development of the orchard and a dramatic decrease in the number of fruit varieties grown. The 1880s mark the turning point in the apparently synchronous development of orchards and their fruits until this time. In this modern era the history of orchards and orchard fruits deviate drastically in growth, with the development of orchards continuing to grow in complexity and the diversity of varieties shrinking rapidly. This era represents the development and standardization of an orchard industry, with a transfer of control from the hands of private individuals and pomologists to the federal government and agribusiness. The work of growing fruit trees is relinquished by small independent farmers to become dominated by professional growers organized in business cooperatives. Scientific changes in orchard management are advanced by the systematic research efforts of universities, and apple growing is fundamentally changed with the discovery of two new varieties, (Red) Delicious and Golden Delicious.

Federal Government Involvement

Several events of the former decade precipitated a more formalized, scientific era in orchard fruit growing in the 1880s. In 1870, the newly formed United States Department of Agriculture (USDA) imported 300 varieties of apples from Russia in the hope of selecting new commercial varieties. Russian apple trees were known to be well adapted to cold winters and springs. If they could be found to have other good commercial characteristics, such as taste and appearance, it was thought they would extend the range of apple growing into the far north of the upper Midwest. The Iowa Agricultural College in Ames cooperated with the USDA in this venture, possibly the earliest example of federal government involvement in fruit development. A decade of field trials yielded only a few varieties with commercial potential.

Government Role in Citrus Development

A more significant event involving the federal government in fruit development occurred in 1874 when the USDA imported a variety of orange from Brazil, and introduced the variety to California. The variety was Navel, a seedless orange that would become the catalyst for the development of a commercial citrus industry in the United States. Prior to the introduction of the Navel orange, orange trees in California were seedlings, derived from the seeds of orange trees in Spanish missions. These "mission oranges" were relatively small in fruit, with soft flesh and many seeds. The Navel orange immediately found favor in Riverside, California, where the first trees were planted (Figures 3.1 and 3.2). In contrast to mission oranges, the seedless Navel orange was found to be large, firm, juicy and highly flavored. The Navel tree had other favorable characteristics too, including

Figure 3.1: *(above)* Historic photograph of citrus orchards in 1915, where the Navel orange was heavily planted, Los Angeles County, CA (from Library of Congress Prints and Photographs Division, courtesy of Walter J. Lumbleau).

Figure 3.2: Historic photograph of one of the original Navel orange trees introduced into Riverside, CA in 1874. The tree was transplanted by President Theodore Roosevelt in 1903 (courtesy of the University of Chicago).

the prolific bearing of fruit and bearing young, as early as the second year after grafting.

The USDA continued to have a dominant role in fostering a citrus industry in the United States. In 1880 the USDA imported the first tangerine plants from the Azores near Portugal, and the first lemon plants from other parts of Europe. The USDA acquired mandarin plants from Australia in 1893 (similar to an orange but smaller and sweeter), and supported the development of citrus varieties for the Gulf States and California, particularly through research at the Florida Agricultural Experiment Station (Klose 50: 77).

Discovery of Fire Blight Organism

In the late 1870s, two more significant events marked the end of the golden era in fruit growing and the beginning of a more rationalized period. In 1878 the causal agent of fire blight was discovered by Dr. T.J. Burrill at the University of Illinois. Burrill's research in plant pathology revealed that fire blight affecting pear, apple, quince, peach, and plum was caused by a single-celled organism, a bacterium, which he called *Bacillus amylovorus* (Figures 3.3 and 3.4). This discovery preceded Robert Koch's development of Germ Theory by several years and was received with a level of skepticism, until 1885 when a series of experiments by Dr. J.C. Arthur confirmed Burrill's

Figure 3.3: *(left)* Watercolor of fire blight damage to a shoot of the Kieffer pear variety, painted in 1889 by USDA artist Miss Mayo. Curiously, Kieffer is considered very fire blight-resistant today (from the USDA 19th and early 20th-century watercolor collection, courtesy of the USDA).

Figure 3.4: *(right)* Watercolor of fire blight damage to the leaves of the Le Conte pear variety, painted in 1890 by an unknown USDA artist (from the USDA 19th and early 20th-century watercolor collection, courtesy of the USDA).

findings. **Bacillus amylovorus**, now called **Erwinia amylovora**, was one of the first plant pathogens to be discovered, emphasizing the economic importance of the disease by this time (Lowther 14: III, 1588). The bacterium was found to be native to the eastern United States, where members of the rose family such as crabapple, hawthorn, mountain ash, and serviceberry were host plants.

Advent of Pesticides

One other event in the 1870s signaled the coming of a scientific age. In 1879 the *Proceedings of the Western New York Horticultural Society Annual Meeting* documented the first use of an insecticide in an apple orchard (Lowther 14: I, 75). The *Proceedings* noted that a western New York fruit grower had sprayed his apple trees with Paris Green dissolved in water propelled by a force pump. The grower aimed to kill canker worm, but had found his ripening apples were also devoid of codling moth larvae. This ground-breaking information was possibly the last of its kind to originate from an amateur organization, such as a horticultural society. In the 1880s the role of pesticide development was formally adopted by the federal government, and most scientific discoveries in fruit growing would then originate with the USDA or an associated system of state agricultural experiment stations.

In 1880 the New York State Agricultural Experiment Station undertook to determine the efficacy of Paris Green. Five years later, the *New York State Experiment Station 1885 Annual Report* confirmed that a highly poisonous bright green powder, Paris Green, a ferrous sulfate, could be used to control codling moth. Paris Green, the first insecticide, was discovered by French horticulturists in the 1870s. France also originated the first fungicide, when in 1882, Bordeaux Mixture or hydrated copper sulfate, was found to treat gray mold or *Botrytis* on grapes, stimulating the development of the French winemaking industry. The USDA first published the formula for Bordeaux Mixture in 1885, and in 1887 recommended that the fungicide be used to treat a number of diseases, including apple scab.

Establishment of Experiment Stations

The role of the federal government in agricultural and horticultural development was institutionalized in 1887 when Congress approved appropriations to establish an agricultural experiment station in every state (the Hatch Act). These federally funded research centers would be associated with university or college campuses, where their first order of business would be pest and disease control for crop plants, including orchards. Gradually, the agricultural experiment stations would also take over the work of orchard fruit breeding and variety selection, and later still would develop partnerships with growers' cooperatives to collaborate in field research. The primary mission of these federal institutions was to increase the profitability of all forms of agriculture and horticulture. Their research efforts were therefore focused on increasing crop quality and yield, and all factors capable of enhancing or reducing yield. Bulletins published by the agricultural experiment stations would become the standard authority for new, valuable information, and through education and regulation the USDA would take the leading role in building a professional industry of fruit growers.

Disease Infestations

Federal intervention in the development of the commercial orchard industry came at a time of the lowest orchard productivity in several decades. The 1880s saw increasingly diminishing orchard yields due to rampant, untreated pest and disease infestations. Many small, independent farmers gave up orcharding, realizing that a successful commercial orchard enterprise required a greater investment of capital, skills, and knowledge than they had at their disposal. The 1880s marked the beginning of the decline in the number of orchards throughout the country, a trend that continued until the end of World War II. The precursor to the trend was pest and disease infestations, though these were somewhat stabilized in the 1890s with the new use of topical pesticides. The trend continued, however, with increasing industrialization and urbanization. The prospect of greater prosperity led more and more farmers to give up their rural lifestyles in favor of industrial jobs in cities. Hired labor costs increased as farm labor shrank, pulled away by the lure of higher paid manufacturing jobs.

Abandonment of Farm Orchards

The 1910 census data reveal that 25 percent of the bearing fruit trees in the United States were lost during the first decade of the 20th century, a decline from 200 million to 150 million trees. (By 1930 the number would decrease to 100 million.) By 1910 only half the farms in the United States had fruit trees, approximately three million farms with an average of 50 trees, or one acre of trees per farm. Editorial pleas in farmers' magazines were common in the years preceding World War I, asking workers to stay on farms as an act of patriotism.

Rise in Commercial Orchards

Coincidentally, the loss of many thousands of orchards through farmers' migration to urban centers paralleled the specialization of the orchard industry. Although half of the fruit trees in the country were lost during the first three decades of the 1900s, the losses were mostly from smaller orchards of five acres or less, (approximately 250 trees for apple and pear orchards), or from farms where a range of crop plants was grown. The orchards that remained were typically larger, more sophisticated operations. Orchard managers of the 1900s would become referred to as "growers" rather than farmers, indicating a distancing from general agriculture. Many of the growers who kept their orchards were willing to incorporate new information in scientific management and specialize in growing orchard fruits alone. As a result they soon began to increase the productivity of their trees.

Influence of Cold Storage

Contrary to a general decline throughout the country, the Pacific Coast states added orchards and increased their numbers of fruit trees during the 1880s and 1890s, and the first decades of the 1900s. The increases were due to the planting of newly settled lands as commercial orchards, which was buoyed by the development of transportation and irrigation systems throughout the western United States, and also by technological advancements in fruit canning. The invention of cold storage technology in the 1890s was a particular boost to the young western orchard industry, which depended on the lengthy haulage of fruits to eastern markets for economic viability. (At the turn of the 20th century, 90 percent of the population lived east of the Mississippi River.) Refrigerated rail cars increased the quality of all western fruits arriving in eastern depots, though especially peaches from California, which were more perishable than most orchard fruits.

The need for cold storage was a necessary outgrowth of the use of pesticides, the specialization of the industry, and a resultant increase in yields, all of which threatened to flood markets with an excess of fruits and therefore depressing prices (Lowther 14: I, 90). Cold storage replaced the 19th century root cellar, enabling growers to hold mass quantities of fruits in a pre-ripened state and spread their release to market over subsequent seasons (Figure 3.5). The earliest cold storage facilities were independently owned businesses located near train depots or shipping centers. By the 1910s the largest growers had formed cooperatives to share

Figure 3.5: Photograph of an early 20th-century fruit cellar built by Mormon settlers in the Fremont River Valley, UT, for orchard fruit storage. Today the cellar is located within the Fruita Historic District of Capitol Reef National Park, UT (S. Dolan 2001).

in the ownership of cold storage buildings. The high cost of mechanical or electrical refrigeration prohibited an individual grower from installing a cold storage facility within their own orchard. However, early 20th century commercial orchards frequently had "common storage" sheds, where fruit was temporarily held after packing (Auchter and Knapp 29: 58). In the western states, packing sheds often had additional space that was used for temporary storage of packed fruit.

Development of Packing

From the 1890s to the 1920s, packing sheds were commonly found in commercial orchards west of the Rocky Mountains, but were more unusual in eastern orchards (Figures 3.6 and 3.7). The greater occurrence of packing sheds in the West was due to a different packing method employed by western growers, but was also indicative of emerging rivalry between eastern and western fruit growers. In the eastern United States, farmers had for several hundred years used oak barrels to store orchard fruits (Figure 3.8). Barrel-making by coopers was a companion industry to

Figure 3.6: *(top)* Early 1900s 'model packing shed' for grading, wrapping, and packing orchard fruit illustrated in a 1920s horticulture textbook (from Auchter 1932, courtesy of J. Wiley and Sons).

Figure 3.7: *(bottom, left)* Photograph of an early 1900s packing shed owned by gentleman farmer Moses Cone in Blowing Rock, NC, to support his extensive orchard operation. Today the historic packing shed and orchards are located within the Moses Cone Estate of Blue Ridge Parkway, NC (S. Dolan, 2001).

Figure 3.8 *(bottom, right)* Historic photograph of coopers making barrels for orchard fruit storage in the 1920s (from Auchter 1932, courtesy of J. Wiley and Sons).

commercial fruit growing until the end of World War II when fruits were no longer stored and shipped in barrels. After World War II, fruits were stored and shipped in cardboard boxes or tote crates moved by forklift trucks.

In the western United States, a lack of suitable hardwood trees precluded the making of barrels, but the abundant softwoods were well suited to the construction of lightweight boxes (Figure 3.9). Fully loaded wooden boxes could not be rolled along like barrels, however, and had to be sufficiently lightweight in order to be lifted. Consequently, fruit boxes were smaller and could hold only one-third of the fruit in a barrel. Therefore, the barrel had an efficiency of scale as a shipping container, and due to its size and shape could be packed quickly and directly beneath orchard trees. As a smaller geometric vessel, the box had to be packed with more care in order to hold the maximum number of fruits. The more time-consuming process of packing boxes meant the activity was performed in a packing shed, shielded from the elements, rather than directly beneath the fruit trees (Figures 3.10 and 3.11). However, western growers turned this apparent disadvantage into an opportunity in an effort to differentiate their produce from eastern growers, whose industry was far better established. The smaller volume of fruit contained within a box meant that the fruit on the bottom was less likely to be crushed or bruised, and the box was a more convenient size and shape for retailers to stack and display open in their stores or fruit stands.

Figure 3.9: *(right)* Historic photograph of initial field packing of apples into boxes in 1909. Note the short trunks of the trees, typical of the period, Larimer County, CO (courtesy of Western History/Genealogy Dept., Denver Public Library).

Figure 3.10: *(below)* Historic photograph of peaches packed in boxes in an early 1900s packing shed, Grand Valley, CO (courtesy of Western History/Genealogy Dept., Denver Public Library, n.d.).

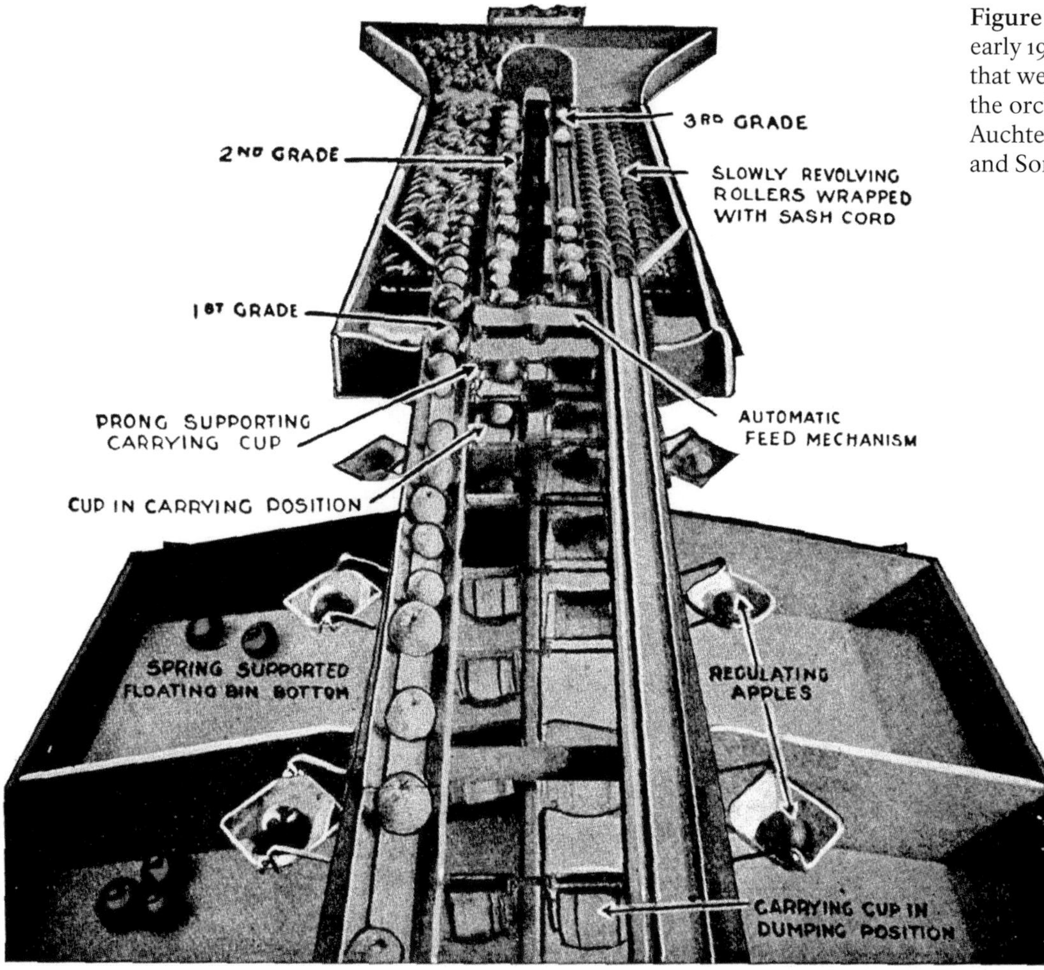

Figure 3.11: Drawings of two, early 1900s fruit-sizing machines that were used to grade fruit in the orchard packing shed (from Auchter 1932, courtesy of J. Wiley and Sons).

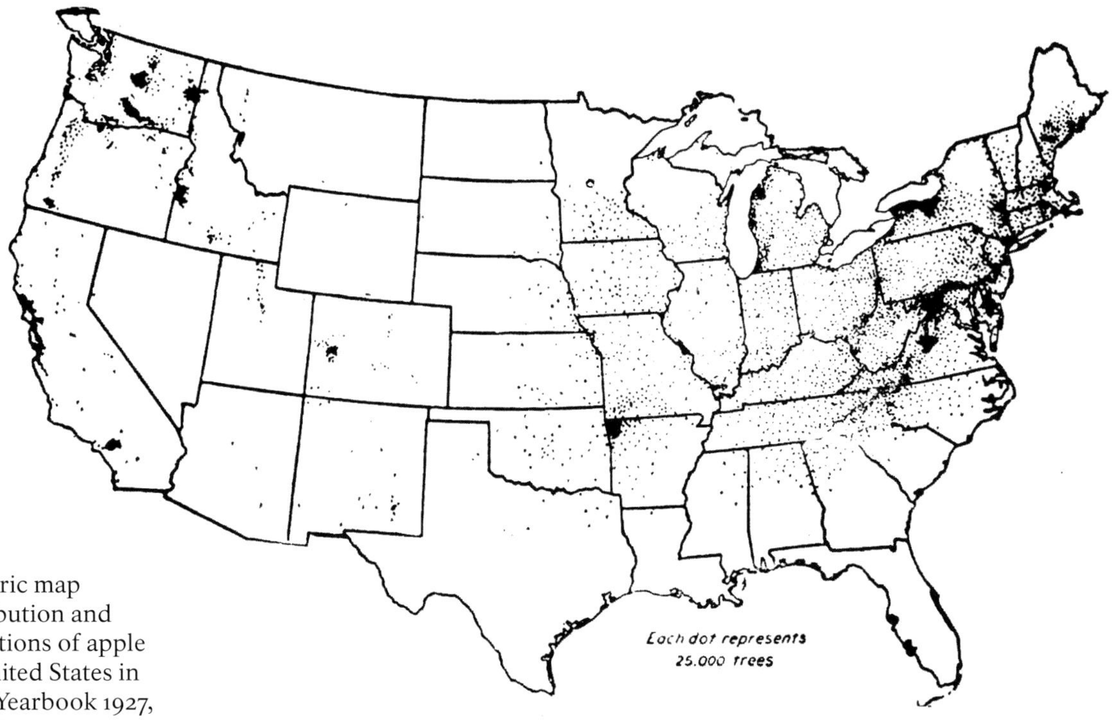

Figure 3.12: Historic map showing the distribution and greatest concentrations of apple orchards in the United States in 1925 (from USDA Yearbook 1927, courtesy of USDA).

Western growers capitalized on the marketing opportunity by developing aesthetic techniques for arranging and tissue-wrapping packed fruit. The importance of packing techniques to fruit quality, and therefore value, is emphasized in the horticultural literature of the early 20th century (Lowther, Auchter and Knapp). The emergence of intense competition between eastern and western apple growers was underscored by a growing concentration of apple orchards in the West (Figure 3.12). Western growers disparaged the quality of fruit packed in barrels and promoted the association of fruit boxes with higher quality fruit through the use of attractive box labels and advertisements. The use of the box rather than the barrel, and its opportunities as a marketing medium was a factor that galvanized western growers and contributed to their earlier development of marketing cooperatives.

Reduction in Variety Diversity

At the beginning of the 20th century, the orchards of the 48 states contained the greatest array of fruit varieties in the country's history. The fervent work of variety development between 1800 and 1880 had vastly increased the numbers of all major orchard fruit varieties. As one example, a USDA bulletin released by the Bureau of Plant Industry in 1904, the *Nomenclature of the Apple: A Catalogue of the Known Varieties Referred to in American Publications from 1804 to 1904*, documented the astounding number of 6,700 apple varieties grown in the United States during the 1800s. But by 1900 changes were afoot in the selection of varieties for new orchards.

New commercial criteria for variety selection were influencing the diversity of fruit trees available from plant nurseries and the range of ripe fruits at retail stands. These criteria were developed by the agricultural experiment stations and were further refined in the decades preceding World War II. They were initially focused on high productivity, and the USDA bulletins recommended lists of the heaviest and youngest-bearing varieties of all orchard fruits. By the end of the 1910, the array of varieties in commercial orchards had narrowed considerably from hundreds of varieties in the 1870s to tens for each species. As the new scientific era proceeded, the criteria for commercial fitness were further refined, and the number of the most important commercial varieties of each species was dropped from dozens to a little more than 10.

Modern Commercial Characteristics

By 1945 the commercial production of all major orchard fruits had come to rely on a handful of varieties that conformed to certain yield-based criteria. These criteria included: abundant fruitfulness or productivity, youthfulness of fruit bearing, more compact size of tree, later blooming (i.e., avoiding frosts and later ripening), pest and disease tolerance or resistance, and commercial fitness of fruit (including the consistency of size, color, and taste of fruits), and tolerance to cold storage and shipping requirements.

The great paring down of varieties created a broad distinction between new orchards of the 1900s and farm and commercial orchards of the 1800s. Whereas orchards in the 1800s often contained several species of fruit and many varieties of each in rows, a commercial orchard in the 1900s often contained only one kind of fruit, with less than four varieties in total, planted out in large, single-variety blocks. New farm orchards of the 1900s were planted with the most common, commercially important varieties rather than connoisseur or local varieties that distinguished the farm orchards of the 1800s (Figure 3.13). Only the number of varieties of

Figure 3.13: Historic photograph of a late 1800s fruit stand displaying many varieties of apples at a county show in the Midwest (from NPS Midwest Regional Office).

citrus fruits and nuts ran contrary to the general pattern of attrition. The commercial growing of citrus and nuts were nascent industries in the 1800s, and their development was greatly intensified in the early 1900s when many new, commercially adapted varieties of orange, lemon, grapefruit (pomelo), walnut, pecan, almond, and hazelnut were created.

Great Reduction in Apple Varieties

The dramatic reduction in the number of varieties grown between 1880 and 1945 is best documented for the apple, as it was the most economically important orchard fruit during this period. Also, the apple was the species with the greatest 19th century diversity, and the greatest 20th century losses. The botanist Liberty Hyde Bailey of Cornell University conducted a survey of plant nurseries in 1892, and found that throughout 40 states, 735 varieties of apple trees were available for sale to growers. The same survey repeated in 1910 by Granville Lowther, a western horticulturist, yielded a 46 percent reduction in the number of apple varieties for sale at nurseries. Robert F. Carlson, a Michigan horticulturist who examined nurseries in 1970, found that only 10 percent of the varieties documented by Bailey were still being propagated—80 years later. An examination of the catalogs published by one of the leading nurseries in the early 1900s, Stark Brothers, revealed a significant reduction in the varieties sold before World War II (Table 3.1). In less than 20 years, between 1918 and 1935 Stark Brothers reduced their number of apple varieties marketed by 80 percent and their pear varieties by 60 percent.

Few Dominant Apple Varieties

By 1910 more than 75 of all apples grown in commercial orchards in the United States were of just 10 varieties. All of these varieties had existed for 80 years or more (Table 3.2). These apple varieties produced fruit abundantly and were popular with the American public. The 10 varieties were not evenly distributed in orchards throughout the country, but were regionally differentiated by their adaptation to certain growing conditions. Baldwin and Ben Davis were grown in far greater quantities than the other top eight varieties. Baldwin was most well adapted to the New England climate, and was the most common apple tree in commercial orchards there and in New York State. Ben Davis was the dominant commercial variety throughout much of the rest of the country (between the 32nd and 42nd parallels), though it was considered to be of highest quality in southern and southwest states. Rhode Island Greening and Roxbury Russet were common throughout the Northeast, Winesap and York Imperial in the South and Midwest, and Jonathan, Northern Spy, King of Tompkins County and Newtown Pippin were widely adaptable, and could be found from Michigan to California.

Reduction in Stark Brothers Nursery Sales of Apple and Pear Varieties 1918 – 1935		
Nursery Catalog Date	**No. Apple Varieties For Sale to Growers**	**No. Pear Varieties For Sale to Growers**
1918	95	29
1928	33	13
1935	19	9

Most Popular Apple Varieties in 1910 (in descending order)	
Apple Variety	**Year of Origination (approx.)**
Baldwin	1750
Ben Davis	1800
Jonathan	1829
Northern Spy	1800
Rhode Island Greening	1650
Roxbury Russett	1650
Winesap	1817
King of Tompkins County (Tompkins King)	1806
Newtown Pippin (Yellow Newtown)	1750
York Imperial	1830

Table 3.1: *(above)* Stark Brothers Nursery catalog data indicating a significant reduction in the number of apple and pear varieties marketed between 1918 and 1935 (source: Stark Bros Nursery Catalogs, 1918-1935).

Table 3.2: *(left)* The top 10 apple varieties grown in the United States in 1910 with their year of origination, in descending order of acreage planted (source: Lowther 14: I, 70).

Within several years nurseries began to promote three other varieties more heavily, and by 1930 the list of the most commonly planted apple varieties had changed a great deal. Two varieties that rose to prominence in orchards were already more than 100 years old, but were just becoming widely popular with growers, as their good commercial characteristics were formally recognized. Cold-hardy McIntosh was selected as a good commercial apple for the Northeast and upper Midwest, where it came to replace Baldwin. Rome Beauty was re-acknowledged as an excellent apple for higher elevations, and was newly planted in the foothills of the Appalachians, Ozarks, Rockies, Cascades, and the Sierra Mountains (Auchter and Knapp 29: 167).

Rise of Delicious Apple

The third apple variety that rose to prominence and even began to top the best-sellers' list in 1928, would become the hallmark of the new period of commercialization and the most commercially valuable variety in the history of apple growing (Carlson 70: 45). The variety's name was Delicious (Figure 3.14). Delicious, which became known as Red Delicious through improved strains, was discovered in Peru, Iowa in 1872, and was first marketed to growers in 1908 (Lowther 14: I, 80). Growers started to recognize Delicious as a valuable new variety in the 1920s when it began to supersede older varieties, particularly Ben Davis (Bright 88: 26). In 1921 Stark Brothers released the first red sport of Delicious, called Starking Delicious. The red sport was the first of several hundred strains of the original variety to be created during the 1900s. A chance mutation on a tree that yielded particularly red fruit, the red sport became a common "improvement" over old varieties of the 1920s, such as Red Rome Beauty, Red McIntosh and Red Northern Spy. Red sports of Delicious turned the originally yellow-red striped skin into a deep red fruit, the color most commonly associated with Red Delicious today (Figure 3.15).

Characteristics of Delicious Apple

By 1925 a Delicious orchard was worth twice the value of a Ben Davis orchard, and a barrel of Delicious could attain the highest price on the market at $6.32 a barrel compared to Ben Davis, one of the lowest prices at $3.54 a barrel (Auchter and Knapp 29: 85). By 1930 Red Delicious was among the five most commonly planted apples in commercial and farm orchards, and growers had dropped Ben Davis from new plantings. By 1942

Figure 3.14: Photograph of the original Delicious variety, the Hawkeye, a red/green apple, borne on a 1908 tree in the former Sonoma State Hospital Orchard at Jack London State Historical Park, CA (C. Cain, 2006).

Figure 3.15: Photograph of the familiar Red Delicious variety developed by Stark Brothers Nursery through red strains (selections) of the original Hawkeye (courtesy of Washington Apple Commission).

Red Delicious' great adaptability to a wide range of growing conditions, its heavy and youthful bearing, its drought tolerance, its blight resistance, and its highly attractive fruit made it the most commonly planted apple variety throughout the country, accounting for 17 percent of all apples produced. Red Delicious' remarkable success would continue to grow over the next 40 years, as described in Chapter 4. Its production peaked in the late 1980s when almost one in every two apples commercially produced in the United States was a Red Delicious (American Fruit Grower 9/97).

Discovery of Delicious Apple

The success of Red Delicious before World War II was not without considerable assistance from the owner of the variety, Stark Brothers Nursery. Founded in Louisiana, Missouri in 1816, Stark Brothers Nursery devoted its propagation and retail efforts during the 1800s to fruit varieties that were well adapted to the Midwest. In 1896 Jesse Hiatt, an Iowa farmer, offered C.M. Stark (president of Stark Brothers) the opportunity to buy the propagation rights to one of his trees, a variety with a broader geographic range. Hiatt had discovered a chance seedling in his apple orchard of Yellow Belleflower 24 years earlier, and had spent the intervening years cultivating the tree and testing its potential in local agricultural competitions.

Hiatt knew he had a winning variety when he offered the tree to Stark in 1896. The variety, which he called Hawkeye, had won first prize in a competition that year (Burford 98, n.p.). Stark Brothers bought the rights to the tree, named it Delicious, and spent $750,000 over the next 20 years on advertising. Stark's marketing strategy included the distribution of more than 8 million "gift trees" to growers as promotional free samples and through farm-to-farm nursery salesmen (Carlson 70: 45). The success of the Stark Brothers' advertising campaign and the excellent commercial characteristics of Red Delicious are reflected in its position at the top of the 1942 list of the most important commercial apple varieties (Table 3.3).

Decline in Pear Popularity

The 20th century trend towards simplification of the number of commercial varieties applied to pears, peaches, cherries, and plums as well as apples, though apples and pears were more dramatically reduced in diversity. Between 1880 and 1945 the pear lost considerable popularity with growers and diminished in commercial importance from second to apple, to sixth in line after apple, peach, plum, orange, and cherry. The pear continued to slip even further after World War II and by the 1980s, accounted for only two percent of the acreage of orchard fruits in the United States, or eleventh in line after orange, apple, almond, pecan, peach, grapefruit, English walnut, plum, cherry, and avocado (Census of Agriculture 80: I, 51).

Two main factors influenced the decline of the pear from commercial production, though one factor was by far the most significant. This was the devastating effect of fire blight, a widespread, fatal disease that remains difficult to manage even today. The other factor was the perishability of the pear and its limited tolerance to lengthy storage and haulage after harvest. The enormous impact of fire blight was to eventually relocate

Table 3.3: The top 10 apple varieties grown in the United States in 1942 with their year of origination, in descending order of acreage planted (source: Carlson 70: 19).

Most Popular Apple Varieties in 1942 (in descending order)

Apple Variety	Year of Origination (approx.)
Red Delicious	1921
Jonathan	1829
McIntosh	1800
Rome Beauty	1848
York Imperial	1830
Stayman Winesap	1895
Northern Spy	1800
Rhode Island Greening	1650
Newtown Pippin	1750
Grimes Golden	1804

the commercial growing of pears from the Northeast and upper Midwest to the semi-arid and arid parts of the Pacific states, where the pathogen was less favored in less humid, cooler temperatures during blossoming. The development of a regional pear industry in Washington, Oregon, and California also had the effect of reducing the number of commercial varieties to those that grew well in those states.

According to Leroy, the 19th century French pomologist, 700 varieties of pear were well regarded in 1867. Marshall Wilder's catalog advertised 70 varieties in 1858, but within 50 years less than 30 varieties were commonly

Most Popular Pear Varieties in 1910 (in descending order)

Pear Variety	Year of Origination (approx.)
Bartlett	1790
Kieffer (Eastern states and South)	1873
Beurre d' Anjou (Pacific states)	1800
Belle Lucrative	1800
Howell	1840
Louise Bonne De Jersey	1780
Seckel	1810
Duchesse d'Angouleme	1808
Flemish Beauty	1810
Garber	1870
Lincoln Coreless	1830
Koonce	1850
Le Conte	1846
Wilder Early	1884
Sheldon	1815
Vermont Beauty (probable synonym Forelle)	1880 / 1700
Vicar of Winkfield	1760
Comice	1849
Guyot	1870
Winter Nelis	1800

Table 3.4: The top 20 pear varieties grown in the United States in 1910 with their year of origination, in descending order of acreage planted (source: Lowther 14: II, 1590).

found. By the 1910s, Bartlett, a French pear introduced in the early 1800s, was more widely grown than all other varieties combined (Figure 3.16). The variety's broad adaptability to a wide geographic range, its heavy and youthful bearing, and ability to self-pollinate (enabling it to be planted as a monoculture), made Bartlett the most popular pear from New York to California. Kieffer, the heat-loving Asian hybrid pear was the second most abundant variety in New York, the eastern seaboard, and throughout the South and Midwest, whereas Beurre d' Anjou, or Anjou, was second to Bartlett in the West (Table 3.4).

Few Dominant Pear Varieties

At the turn of the 20th century, a new, late-ripening strain of Bartlett called Winter Bartlett was becoming popular with growers, and two other varieties, Beurre Bosc (Bosc) and Doyenne du Comice (Comice), were gaining popularity in the Pacific Northwest (Figures 3.17 and 3.18). In 1930 the state of New York still held onto its century-long status as the top pear producer, though by this time it shared this status with California. A number of less commercially valuable pear varieties could still be found throughout the Northeast and Midwest, but by this time, California grew Bartlett pears almost exclusively (Figure 3.19). By World War II, Anjou and Bosc would become relatively more important in the Pacific Northwest, and the stage was then set for the post-war distribution and dominance of the three most important commercial pear varieties in the late 20th century: Bartlett, Anjou, and Bosc, in California, Washington, and Oregon.

Figure 3.16: *(left)* Watercolor painting of Bartlett pear variety (from the USDA 19th and early 1900s watercolor collection, courtesy of the USDA).

Figure 3.17 *(right)* Watercolor painting of Winter Bartlett pear variety (from the USDA 19th and early 20th-century watercolor collection, courtesy of the USDA).

Figure 3.18: Watercolor painting of Doyenne du Comice (Comice) pear variety (from the USDA 19th and early 20th-century watercolor collection, courtesy of the USDA).

Figure 3.19: *(below)* Historic map showing the distribution and greatest concentrations of pear orchards in the United States in 1925. The quantity of orchards depicted by each dot is unknown (from USDA Yearbook 1927, courtesy of USDA).

Dominance of Elberta Peach

Figure 3.20: Watercolor painting of Elberta peach variety (from the USDA 19th and early 20th-century watercolor collection, courtesy of the USDA).

In the modern era of orchard fruit growing, peach orchards would become dominated by one variety rather like pear orchards. Like the Bartlett pear, the 19th century American peach variety Elberta became the most abundant and ubiquitous commercial variety by 1910 (Figure 3.20). Unlike pear orchards however, peach orchards retained more diversity throughout the 1900s, each typically having several other varieties in addition to Elberta. The larger diversity of peach varieties was partly due to the great perishability of the peach, promoting a greater reliance on the canning industry and a dependence on the traditional, local marketing of peaches at truck stands. A diversity of varieties was also promoted by the faster and shorter lifecycle of the peach, only 15 to 30 years versus 80 to 150 years for apple and pear, and the greater tendency of the peach to hybridize. Consequently, peach orchards were renewed more frequently with new trees, and other varieties could be incorporated.

Ubiquity of Peach Growing

At the turn of the 20th century, peaches were grown in three-quarters of the 48 states and were abundant in all but the most northern latitudes of the upper Midwest. The tempering climatic influence of the Great Lakes enabled growers to plant peaches in Ohio, Illinois, and Michigan. Many excellent clingstone (canning) and freestone (fresh market) varieties existed throughout the country, which were often quite narrowly adapted to a small region of origin. Despite the dominance of the hardy Elberta throughout the country, local varieties held their popularity up until World War II because they extended the season before and after Elberta's harvest and provided best quality when hauled only a short distance (Table 3.5). The state of Georgia grew the most peaches in 1910, a status it relinquished to California by the 1930s.

In the early 1900s, California's enormous Central Valley became an industrial powerhouse of fruit production, challenging the fruit-growing economies of many southern and Midwestern states. With the boom in peach growing in California, the peach-growing districts of Ohio, Illinois, and Michigan were among the first to shrink. The Great Depression of the 1930s saw an even greater rate of agricultural development in California than in the first decades of the 1900s, setting the stage for California to become one of the world's largest fruit producers after World War II (Figure 3.21).

Most Popular Peach Varieties in 1930 (in descending order)

Freestone Variety	Clingstone Variety
Elberta	Phillip
Lovell	Tuscan
Salwey	Palono
St. John	Peaks
Early Crawford	Orange
Late Crawford	Hauss
Belle	Heath Cling
Carman	Early Wheeler
Levi	
J.H. Hale	
Triumph	
Muir	

Table 3.5: The top peach varieties grown in the United States in 1930, in descending order of acreage planted (source: Auchter and Knapp 29: 171).

Figure 3.21: *(below)* Historic map showing the distribution and greatest concentrations of peach orchards in the United States in 1925. The quantity of peach orchards depicted by one dot is unknown (from USDA Yearbook 1927, courtesy of USDA).

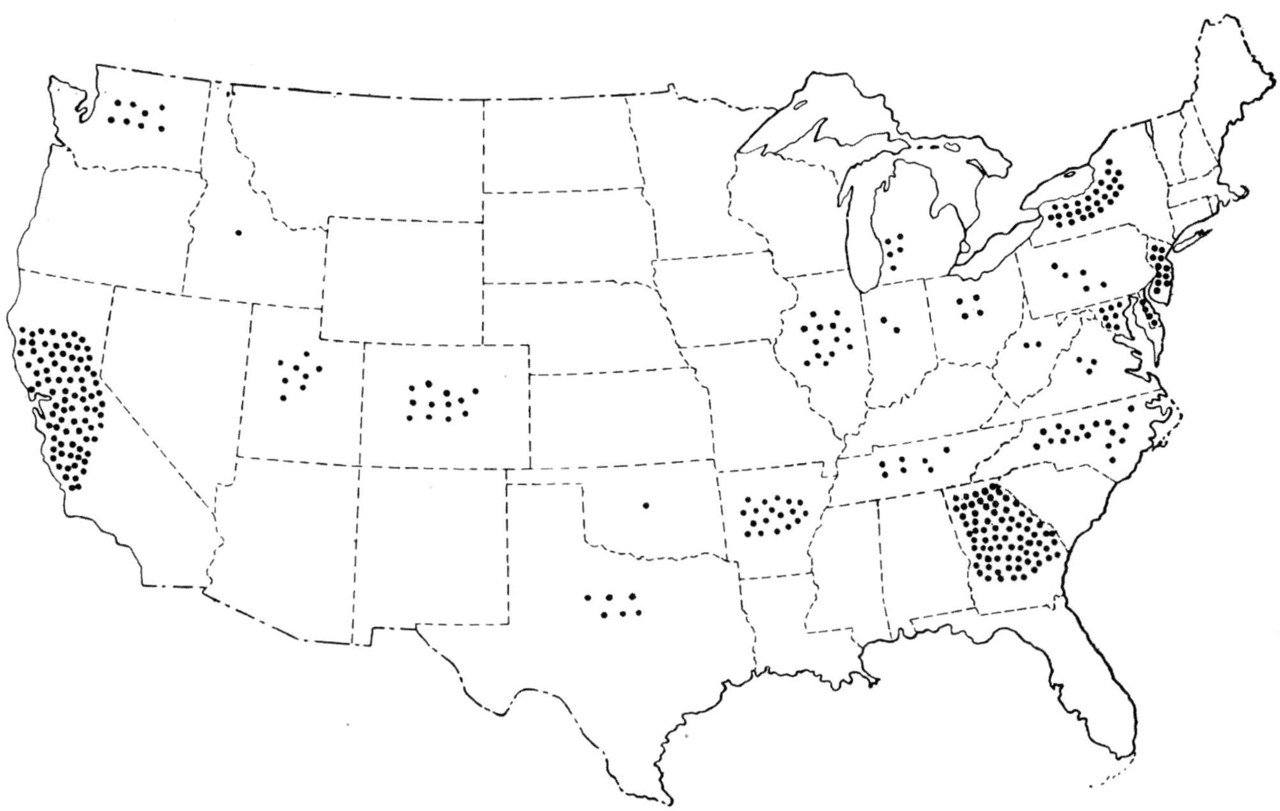

Regionalization of Cherry Growing

Contrary to the relative ubiquity of peach orchards, commercial cherry growing became highly regional in the early 1900s. The main regions of cherry growing were divided in two according to the different climatic adaptations of the two important cherry species. The hardier sour cherry varieties of *Prunus cerasus* (Cherry plum) were mostly grown in the upper Midwest, while the less hardy sweet cherry varieties of the species *Prunus avium* (Mazzard cherry) became centered in the Pacific Northwest.

Montmorency was by far the most common sour cherry variety and Bing was the most important sweet cherry. Sour and sweet cherry trees could still be found east of the Rockies on midwestern and eastern farms, though in far fewer numbers than in the new commercial orchards of Ohio, Michigan, Wisconsin, and Iowa for sour cherries, and in Washington and Oregon for sweet cherries. Some sweet cherry production held on in western New York and western Michigan, where some of the oldest varieties were still grown (Table 3.6). Most of the varieties were European, though curiously, the Bing variety was developed in Oregon by the younger brother of Henderson Luelling, the nurseryman who hauled hundreds of fruit trees along the Oregon Trail in the mid 19th century. Seth Luelling developed Bing in the 1880s, naming the variety for a loyal Chinese employee who had served him for many years.

Alaska Experimentation with Crabapples

During the early 20th century, the Alaska Experiment Station at Sitka experimented in growing sour cherries as a potential orchard fruit for Alaska, and met with limited success. Today, some sour cherry varieties have a minor place in Alaska's small commercial orchard operations.

Table 3.6: The top cherry varieties grown in the United States in 1930, in descending order of acreage planted and by geographic region (source: Auchter and Knapp 29: 173).

Most Popular Cherry Varieties in 1930 (in descending order)

Sour Cherry Variety	Sweet Cherry Variety	
	Pacific Northwest	***W. New York and W. Michigan***
Montmorency	Bing	Black Tartarian
Early Richard	Lambert	Windsor
English Morello	Napoleon	Schmidt
Ostheim		Napoleon
Dyehouse		Yellow Spanish
Large Montmorency		

However, extremely cold-hardy Russian and Canadian apple varieties have become the most important orchard fruits in the state (Wheeler 7/99). The Alaska Experiment Station also explored the use of crabapples for the state's orchards in the early 20th century, due to their extreme cold hardiness. Crabapples are the common ancestors of domestic apples, as members of the same genus. Their fruit is smaller however, typically 1½-2 inches in diameter, and very tart to taste. Crabapples are more dwarf in stature than standard domestic apple trees, and have more of the wild characteristics of seedling apple trees, often with twisted, gnarled and thorny branches. Since the early 20th century, however, the horticultural development of crabapples has exploited the ornamental value of the tree rather than edible fruit. A garden stature and beautiful blossom and fruit clusters have led the development of crabapples into the landscaping arena rather than for orchard fruits.

Regionalization of Plum Growing

Plum growing also became a more regionally specialized commercial industry in the early 1900s, though no single variety dominated in popularity. During the 1800s plum growing evolved as it spread from the eastern to the western United States. In the 18th and early 19th centuries, the eastern United States grew an array of varieties of the European and Damson species *Prunus domestica* and *P. institia*, though most plum trees were grown in fruit gardens rather than orchards (Hatch 98: 107). However, the European and Damson varieties were not well adapted to the hot summers of the south or the severe winters of the Midwest and this limited their geographic range. In regions where European or Damson plums would not grow well, varieties of native plum species were developed, particularly from the two species *Prunus americana* and *P. angustifolia*.

Native Plum Varieties

The discovery of Miner plum in 1814, the first variety developed from a native species, *Prunus angustifolia*, stimulated the creation of many more American varieties during the 19th century. The majority of these varieties were developed and grown in the south and Midwest, where they became part of small commercial orchards that produced plums for jams and jellies. Among the most important native plum varieties developed in the south after Miner were Wild Goose in Tennessee (1850) and Robinson in North Carolina (1879). In the Midwest, several important varieties were developed in Iowa, namely Wolf (1850), Quaker and Forest Garden (1862), and Weaver (1875). Other important native plums developed in the Midwest were Rollingstone in Minnesota in the 1850s, and De Soto and Cheney in Wisconsin in the 1860s (Hedrick 98: 170).

Japanese Plum Variety Development

A further evolution in plum growing was spurred by the introduction of Japanese plums in 1876. The development of the Japanese plum, *Prunus triloba*, was largely undertaken by Luther Burbank, a California horticulturist who worked extensively on the development of new *P. triloba* varieties during the 1880s and 1890s. Burbank's name is associated with the improvement of many Japanese plum varieties, including Burbank and Satsuma, and with the creation of many Japanese-American hybrid plum varieties, such as Gold, Splendor, and Wickson (Harwood 14: 128). Japanese plums and their hybrids are distinguished from European plums by their round, rather than oval (or egg-like) shape.

By 1910 California had more plum orchards than any other state, and the Pacific states produced 78 percent of the country's plums (Lowther 14: III, 1692). A combination of many European and Japanese plum varieties were grown in the Pacific states, where they produced fruit for the fresh and processed markets. A range of American plum varieties remained dominant in the south and Midwest, and were used primarily for processing. The growing of European plums had declined considerably in the Northeast due to the ravages of plum curculio, an insect pest that had devastated older plum growing areas in the 19th century (Table 3.7).

Orchard Characteristics of Period

Between 1880 and 1945, the diversity of orchard fruit varieties was honed to a relatively small number that were most commercially viable, and the overall appearance of orchards changed considerably. The field research of the agricultural experiment stations yielded scientific information that was manifest by 1910 in new orchard management techniques. The visible influence of the new horticultural science was a change in the form, shape, and layout of orchard trees. The new techniques held sway until after World War II when another era of orchard management emerged. One of the greatest changes to the appearance of early 20th century orchards was in the form of individual fruit trees. In the 19th century, fruit trees had tall, unbranched trunks generally five or more feet in height; trees were generally unpruned and their form was created by browsing livestock and deer. By the early 20th century, fruit trees were "low-headed" with a trunk just 18-36 inches high before the emergence of the first limbs (Figures 3.22 and 3.23). E.C. Auchter, a principal horticulturist of the USDA, explained the change in philosophy in 1929:

> *The region of the trunk of the tree where most of the main framework branches start is spoken of as the "head" of the tree...A few years ago trees were often headed high, 5 or 6 feet from the ground. It was felt that the ground under such trees could be cultivated more thoroughly and easily. In recent years, however, trees are being*

Most Popular Plum Varieties in 1930 (in alphabetical order)		
European Variety *(Prunus domestica and P. institia)*	**Japanese Variety** *(Prunus triloba)*	**American Variety** *(Prunus americana and P. angustifolia)*
East and West Coast	***West Coast***	***Midwest and South***
Arch Duke	Abundance	Ames
Bavay	Burbank	America
Bradshaw	Chabot	Cheney
Damson	Climax	De Soto
German Prune	Duarte	Forest Garden
Grand Duke	Formosa	Golden
Green Gage (Reine Claude)	Gaviota	Hawkeye
Italian Prune (Fellenberg)	Kelsey	Pottawattamie
Imperial Epineuse	Red June	Wayland
Lombard	Santa Rosa	Whittaker
Monarch	Satsuma	Wild Goose
Moore Arctic	Wickson	Wolf
Pacific		Wyant
Pond		
Quackenboss		
Shropshire Damson		
Tennant		
Tragedy		
Yellow Egg		

Table 3.7: The top plum species and varieties grown in the United States in 1930, in alphabetical order and by geographic region (source: Auchter and Knapp 29: 174).

headed much lower, 20 to 36 inches. . .It is plainly evident that lower headed trees can be pruned and sprayed, and the fruit thinned and picked more easily and cheaply. Cultivating tools with extension sides now permit the orchardist to work the ground satisfactorily and conveniently under the branches. Less whipping about and bending by the winds is experienced with low-headed trees, (Auchter 29: 210).

Figure 3.22: *(right)* Historic photograph of a "model low-headed orchard" in a 1914 horticulture textbook, showing apple trees with short trunks and an open-bowl pruning style, planted in 1896 (from Lowther, 1914, courtesy of the Encyclopedia of Horticulture Cooperation).

Figure 3.23: *(below)* Historic photograph of a Colorado State prize-winning orchard in the 1910s, showing cherry trees with short trunks and an open-bowl pruning style (courtesy of Western History/ Genealogy Dept., Denver Public Library).

Low-Headed Trees

The emergence of the technique of pruning for short trunks, or low-heading, was the beginning of the widespread attempt to control the size of the fruit tree, a practice that would greatly intensify after World War II. A shorter trunk lowered the entire canopy by a few feet, making it more accessible for management Figures 3.24, 3.25, 3.26 and 3.27). Shortening the trunk of the tree also caused the tree to bear fruit earlier by reducing the hormonal effect of apical dominance (a phenomenon involving the dominance of the apex or apical bud suppressing the development of lateral axillary buds). Coincidentally with the technique of low-heading, the practice of grazing livestock within orchards was abandoned, and fences or hedges were erected to exclude livestock and wildlife. Poultry were often still permitted to browse within orchards, as the nutritional potency of their manure was well recognized.

Figure 3.24: *(top)* Photograph of low-headed apple trees in Moses Cone Estate Flat Top Orchard planted in 1899, Blue Ridge Parkway, NC (S. Dolan, 2001).

Figure 3.25: *(bottom, left)* Photograph of a low-headed Baldwin apple tree in Flat Top Orchard, planted in 1899. Characteristically, the tree was "headed" in its first or second year to form a short trunk and then pruned into an open bowl style, Moses Cone Estate, NC. (Barry Caldwell, Park Exhibit Specialist, for scale (S. Dolan, 2001).

Figure 3.26: *(bottom, right)* Photograph of the low-headed Mulford apricot orchard planted in the 1920s, Fruita Historic District at Capitol Reef National Park, UT (S. Dolan, 2001).

Figure 3.27: Photograph of a low-headed cherry orchard planted in the 1920s, Fruita Historic District at Capitol Reef National Park, UT (S. Dolan, 2001).

Advent of Synthetic Fertilizers

However, the practice of grazing chickens in the orchard to provide manure was phased out during this period as growers realized the greater efficacy of synthetic fertilizers. The identification of nitrogen, potassium, and phosphorous as plant macronutrients, and the discovery of the role of nitrogen-fixing bacteria in converting atmospheric nitrogen to soluble nitrate, led to the widespread use of synthetic fertilizers and the frequent practice of cover-cropping. Following scientific guidance from the USDA agricultural experiment stations, growers planted root-nodule forming leguminous plants, such as alfalfa, buckwheat, peas and vetch as orchard groundcovers, and incorporated manure periodically rather than on an ongoing basis.

Adoption of Pruning Styles

In addition to the shorter trunk form of fruit trees in the early 1900s, new pruning techniques also changed as a result of the development of pruning skills by the common grower. Pruning styles were generally adopted for all types of orchard fruits, including citrus and nuts, with most fruit trees being trained into one of two dominant shapes: the "central leader (pyramidal) style" or the "open bowl (vase) style." Both styles were implemented on a low-headed trunk (Figure 3.28). In the central leader style, the main vertical shoot was trained to develop a scaffold of well-spaced, nearly horizontal branches, rather like a ladder. The horizontally borne branches intercepted more sunlight than untrained branches, which increased blossoming, and the crotches of horizontal branches were stronger when laden with fruit. In the open bowl style, the central leader was removed and three to five main shoots were permitted to radiate from the head of the trunk, rather

Figure 3.28: Diagram of three tree pruning styles popular in orchards in the 1881-1945 period. Of these, the central leader and open bowl styles were most common. The modified central leader style was more popular after World War II (based on Tukey, 1964, courtesy of Cornell University Press).

like an inverted tripod or tetrapod. A bowl-like scaffold was created with an open center, allowing sunlight to penetrate deep into the canopy.

One of the major advantages of the open bowl style of pruning was greater control over the height of the tree than with the central leader style, though one important disadvantage was the potential for a weaker system of more acute crotches for bearing fruit. Both styles were perceived to have strengths and weaknesses, and both were greatly favored over the "natural" or unpruned style of the 19th century, which bore less fruit and had a more unwieldy canopy. The central leader style was generally more common, as it required less skill and labor than the open bowl style, which was achieved through more frequent pruning interventions.

Orchard Spacing in Period

Another change in the appearance of early 20th century orchards was the move towards a wider spacing layout. The low-heading of trees and the acquisition of tractors by growers led to a need for greater turning radii between the rows of the more tightly spaced orchards, such as peach and plum. The need for greater maneuverability became manifest in wider spacing between rows of trees, rather than within rows, and peach and plum orchard spacing was frequently changed to a rectangular rather than a square spacing layout (Figures 3.29 and 3.30). Trees planted at 15 feet within rows and 20 feet between rows became common, rather than the more typical square signature in the 1800s. In addition, new apple and pear orchards were also planted at a wider spacing as a result of the low-heading of trees and a new belief in greater fruit yields in mature trees with wider spacing.

During the first decade of the 1900s, the New York Agricultural Experiment Station at Geneva influenced growers to plant their apple and pear trees at a wider spacing by disseminating data on orchard spacing and fruit yield.

Figure 3.29: *(above)* Historic photograph showing rectangular spacing in a low-headed apple orchard in the 1910s. Spacing between the rows is wider than within the rows. Smudge pots of burning kerosene repel codling moths during fruit set, near Cañon City, CO (courtesy of Western History/Genealogy Dept., Denver Public Library).

Figure 3.30: *(right)* Historic photograph showing wide, rectangular spacing in a low-headed orchard to allow room for mechanical cultivation, 1920s (from Auchter 1932, courtesy of J. Wiley and Sons).

Increase in Apple and Pear Tree Yield with Increased Tree Spacing (1910)	
Tree Spacing	**Yield (Bushels/Acre)**
	1 bushel = 4 pecks = 42 pounds
30 x 30 feet	186
31 x 31 feet to 35 x 35 feet	222
36 x 36 feet to 40 x 40 feet	229

Table 3.8: Early 1900s USDA research data on the relationship of orchard spacing and fruit yield, indicating greater yield with greater spacing (source: Lowther 14: I, 158).

Their research indicated that apple and pear orchards of the 1800s had been planted too close at 20- to 30-foot squares, and this compromised the yield of fully grown, mature trees. It was thought that the largest of the most important apple varieties—Baldwin, Rhode Island Greening, and Northern Spy—could be planted at 50 feet apart, and that regular-sized McIntosh, Jonathan, and Rome Beauty could be planted at 35 feet apart (Table 3.8).

Advent of Filler Trees

To compensate for the inefficient use of space during the early lives of widely planted apple and pear trees, "filler trees" were now added. The concept of filler trees was conceived in the early 1900s as a way to optimize the use of land within the orchard in between the wide spacing of "permanent" trees (Figure 3.31). Filler trees were most commonly inserted within a square spacing of 40 feet to form what was called a quincunx system. A filler tree of a youthful bearing, smaller variety was located at the center of a square of four trees, achieving a spacing of 28-¼ feet. The yield from filler trees supplemented the grower's income until the permanent trees occupied their share of the space, a period of 10 years or more. In theory, the filler trees were then removed. Filler trees could be of the same variety as the permanent trees, but were more often a faster growing, younger-bearing variety, such as Wealthy, Wagener, Duchess, or Missouri Pippin.

Filler trees were regarded as having both advantages and disadvantages, and were not as common as single-aged orchard blocks of varieties in squares or rectangles. The advantages were early economic gains and beneficial services in cross pollination, a factor recognized by the late 1920s. The disadvantages were the limitations imposed on tractor access for cultivation and spraying, and their tendency to retard the growth of the permanent trees, when filler trees remained for too long. The decision to remove the filler tree was a hard call to make; the temptation was to leave them for "just another year" (Auchter and Knapp 29: 153).

Figure 3.31: Historic photograph of a cherry orchard with inter-planted young filler trees in Canyon County, ID, 1941 (courtesy of Library of Congress Prints and Photographs Division, Farm Security Administration, Office of War Information).

Improvement of Irrigation Systems

Between 1880 and 1945, irrigation systems became a typical feature of western orchards on arid or semi-arid lands, but were rare in orchards east of the Mississippi. American Indians built the first irrigation systems in the country's desert southwest, long before the arrival of the Spanish and their development of irrigated mission orchards. As the Spanish missions were being secularized, Mormon settlers were digging irrigation ditches in Utah to water their newly planted orchards. The emigrants who crossed the Oregon or California Trails to the Pacific states often adopted abandoned placer miners' ditches to irrigate their homestead orchards. As the new era began, the first irrigation systems of western commercial orchards were hand-dug ditches, often built cooperatively by a group of settlers, or their hired Chinese laborers. The earliest irrigation systems were crude; water was diverted from a river or creek by a dam of piled rocks or logs, and delivered via a series of open earthen ditches that dissipated a sheet of floodwater across the orchard floor (Figure 3.32).

As the period progressed, the federal government and private corporations became involved in the business of irrigating the western states, and systems became increasingly sophisticated. By the turn of the 20th century, growers were lining their irrigation ditches with concrete and lacing their orchards with a system of secondary ditches running in between tree rows to deliver water to each tree. The pattern of water circulation and rate of flow was controlled by a series of head flumes, consisting of wooden gates that could be slid up and down within concrete braces. A less frequent variation on this common system was the use of clay tile rather than concrete, to line the ditches.

Figure 3.32: Historic photograph of an unlined, hand-dug irrigation ditch watering an orchard and pasture in the 1910s. Note the fence to exclude grazing cows from the low-headed orchard, CO (courtesy of Western History/ Genealogy Dept., Denver Public Library).

Government Role in Irrigating West

By the late 1910s growers were enclosing irrigation ditches to conserve water (Fortier 10: 6). The material of the delivery medium gradually changed from concrete to steel, and by the 1930s many large, piped irrigation systems were powered by pumps. A series of congressional acts in the late 19th and early 20th centuries, such as the Carey Act and the Reclamation Act, permitted the federal government to intervene in watering the West. The federal government established partnerships with state governments and private corporations in North Dakota, South Dakota, Kansas, Nebraska, Oklahoma, New Mexico, Arizona, California, Nevada, Utah, Colorado, Wyoming, Montana, Idaho, Washington, and Oregon to build dams and reservoirs for the diversion and storage of water, and to build canals and ditches for land reclamation. By 1913, 8,000 miles of canals and irrigation ditches had been constructed by the Reclamation Service (within the Department of Interior), along with five million acre-feet of reservoirs—enough water to irrigate 1,300,000 acres of land. (An acre-foot is a volume measurement of one acre area with a depth of one foot.) (Figure 3.33) (Lowther 14: II, 1164).

The involvement of the federal government in irrigating western lands continued through the Great Depression of the 1930s when the largest engineering works were implemented. Among the enormous impacts of these long-term federal investments was the agricultural development of the West on an industrial scale. Large commercial orchards in California's Central Valley and in the arid eastern sides of Washington and Oregon states were the primary beneficiaries of these reclamation projects.

Figure 3.33: Historic photograph of the concrete-lined Orchard Mesa Irrigation Ditch in Mesa County in 1911, an example of a federal Reclamation Service project, CO (courtesy of Western History/Genealogy Dept., Denver Public Library).

Advent of Fruit Standards

Other federal regulations enacted in the Progressive Era brought changes to commercial orchards. New laws for consumer health and safety and the protection of the agricultural industry resulted in higher standards for orchard management. Standard grades for apple sizes and quality were adopted in the 1910s, which improved the reputation of barrel apples. In the West, standards were tightened for boxed apples, requiring packers to be licensed for the handling of fruit and to possess their own rubber stamp for packed boxes. New quarantine rules affected the interstate shipment of fruit, causing impromptu inspections at border crossings and fines for growers and distributors who attempted to ship infected fruit.

Fruit cleanliness became an even greater concern for growers in the late 1920s, when the newly formed U.S. Food and Drug Administration required that all fruits be free of lead arsenate residue before retail sale. By the 1920s, lead arsenate was the pesticide of choice for the control of codling moth on apples. Although its efficacy was limited, control was improved through repeated applications. Apple trees were sprayed up to 10 times during one season to prevent adult moths laying eggs within the developing fruit. Consequently, a typical commercial apple was coated with a thick gray-white residue by the time it reached the consumer.

In 1926 the British government issued a formal complaint about lead arsenate residues on apples imported from the United States, and the federal government responded with a law requiring the cleaning of apples before sale (Bright 88: 45). The impact of the new law on the grower was considerable. Growers no longer packed fruit in their own orchards, but shipped their yield to a central processing center that could handle bulk washing, sorting, grading, and packing. Raising the quality standards in this manner led to the disappearance of the packing shed from western orchards. During the 1930s packing sheds gradually became obsolete and were superseded by cooperatively owned, centralized washing and packing facilities.

20th Century Pest Control

The control of codling moth remained a severe problem for apple growers until after World War II, when the new organochlorine insecticide DDT became widely distributed. The use of pesticide sprays was considered requisite to a successful orchard enterprise by 1910, and for the first several decades of the 20th century, several broad-spectrum toxins replaced the use of Paris Green and Bordeaux Mixture. Lime sulfur superseded Bordeaux Mixture for the treatment of apple scab, scale, and fungi, and nicotine was used to kill aphids. Kerosene oil emulsion and poison baits were used against a range of pests, though of all the pesticides, lead arsenate, used to control biting insects, was the most hazardous and persistent.

Pest Control Operations

Pesticide sprays were delivered by spray rigs, and powders were emitted by dusters, both drawn by horses or tractors moving slowly up and down tree rows (Figures 3.34 and 3.35). Spray rigs were large pieces of machinery requiring a large wood stave tank for storage of the liquid in addition to a pump to propel the spray into a hose, rod, and nozzle. Dusters required a smaller hopper rather than a tank to contain the powder, and were therefore more portable. However, both sprayers and dusters were relatively bulky and were mounted onto a truck base, bearing a wood or steel truss tower to support an operator. One operator was stationed in the tower and aimed a 10-foot-long sprayer rod towards the top of the tree canopy, while another stood at ground level and aimed the spray up into the tree from below. Another man drove the tractor or controlled the speed of the horse.

The invention of the spray gun in the late 1920s replaced the heavy steel rod and nozzle of sprayers and dusters. The hand-triggered spray gun was mounted directly at the end of a hose, and could propel fluids 30-40 feet into the air. Spraying an acre of orchard remained a day's work for several men, however, and required the slow, meticulous delivery of several hundred gallons during the course of one application. During heavy predations by codling moths, growers would organize their laborers

Figure 3.34: Historic photograph of spraying in an orchard in 1911 using a horse-pulled spray tank and long spray rods, to reach the tree tops, Grand Valley, CO (courtesy of Western History/Genealogy Dept., Denver Public Library).

Figure 3.35: Historic photograph of spraying in an orchard in 1935 using a truck-mounted spray tank and powerful spray gun, Hightstown, NJ (courtesy of Library of Congress Prints and Photographs Division, Farm Security Administration, Office of War Information).

to spray at night to have the greatest impact on the night-flying insects. Repeated applications were needed due to the contact, rather than systemic activity of the various sprays. A large volume of pesticide was needed to coat the large orchard trees, and wide orchard spacing was needed to accommodate the trees and the bulky equipment for pest control. The intensely laborious nature of pest control was one of the critical issues that led to a revolution in tree size and orchard management after World War II. However, until 1945, the typical orchard had a distinctly large scale, and a very challenging pest control regimen.

Adoption of Delicious Seedling Rootstocks

During the 1930s and 1940s, the commercial success of the Red Delicious apple variety caused a change in the nature of apple tree propagation by plant nurseries. During the 19th century and for the first decades of the 20th century, the scions of apple tree varieties were grafted onto the roots of any seedling stock that was readily available. As Red Delicious became recognized as a broadly adapted and versatile variety in a broad range of climatic, soil, and moisture conditions, its widespread planting led to the use of its seeds to raise trees for rootstocks. Nurseries obtained the seeds of Red Delicious from apple processing plants where the fruit was crushed or milled for juices or baked goods. Between the late 1930s and the mid-1950s, most varieties of new apple trees were grafted onto the roots of Red Delicious seedlings, which increased the adaptability of the most important commercial varieties.

Figure 3.36: Photograph of Golden Delicious apple variety (courtesy of Washington Apple Commission).

After World War II, Red Delicious apples were sold throughout the world, and their seedling progeny became the most widely used source for seedling rootstocks. Today, many less developed countries continue to grow the majority of their apple varieties on Red Delicious seedling rootstocks. The seeds of Red Delicious are obtained from fruit processing industries in those countries, a statement about the ubiquity and economic importance of the variety.

Rise of Golden Delicious Apple

After World War II, the Golden Delicious apple variety became the almost equally successful yellow counterpart to Red Delicious (Figure 3.36). Golden Delicious would become second only to Red Delicious as the most planted apple variety. The success of Golden Delicious was slower arriving than Red Delicious, however, even though both varieties emerged between 1880 and 1945. By the end of this period Golden Delicious was growing in popularity, but accounted for only 1.8 percent of all apples produced, and ranked eighteenth on the list of the most important commercial varieties (Carlson 70: 34). During the 1950s and 1960s, Golden Delicious was widely planted and moved into the number two position behind Red Delicious for the last three decades of the 1900s.

Discovery of Golden Delicious Apple

The two most popular apple varieties of the 20th century were related by their names and commercial successes only. Golden Delicious was discovered in the 1880s, slightly later than Red Delicious, as a chance seedling in an orchard of two yellow apple varieties, Grimes Golden and Golden Reinette, in Clay County, West Virginia. No doubt the owner, Anderson Mullins, was familiar with Jesse Hiatt's 1893 deal when he offered his propagation rights to Stark Brothers in 1914. Twenty years had passed and now Paul Stark was president of the nursery corporation. While little information is available on Hiatt's price for Red Delicious, the value of the Golden Delicious deal is widely published, probably because the price was significantly higher by this time. Stark Brothers purchased the propagation rights to the so-called "Mullins Seedling" for $5,000 and immediately installed an iron cage and alarm system around the tree. A more competitive market environment and 20 years of inflation had perhaps taught Stark Brothers to more heavily protect their new investment.

Characteristics of Golden Delicious

Stark Brothers Nursery immediately recognized that Golden Delicious had some exceptional commercial qualities. The variety was extremely adaptable to a range of climates, bore fruit young and abundantly, and had the additional virtue of producing lots of pollen over an extended blooming period, making the variety a good pollinator for other apple varieties. In fact the variety was the perfect pollinator for Red Delicious, a variety whose only apparent weakness was its inability to self-pollinate, requiring a companion variety in the orchard to achieve good fruit set. Red Delicious' "self-unfruitful" characteristics went generally unheeded until the late 1930s when the variety had become so singularly planted that pollination was very poor and yields were extremely low. Growers began to regard Golden Delicious as the perfect antidote to Red Delicious' cross pollination need, and in 1940 Stark Brothers Nursery Catalog prescribed a combination of Red and Golden Delicious as the "model apple orchard" (Figure 3.37).

Influence of the Great Depression

The interplanting of Red Delicious with Golden Delicious was also promoted by the Washington Apple Commission, the first apple marketing corporation formed in the United States. Founded during the Depression years as a mechanism for counteracting severe economic impacts on the industry, the Washington Apple Commission took on the promotion of Golden Delicious as a pollinator and a money-maker for the future. The economic benefits of the Commission to Washington growers were scarcely realized until after World War II, however, when Washington State became the fastest growing apple region of the country. The Great Depression years brought many challenges to growers, as it did to all farmers, and many commercial orchards were abandoned. In the words of A.C. Bright,

Figure 3.37: Photograph of the Chestnut Orchard in the Fruita Historic District with a mix of Red and Golden Delicious apple trees planted before World War II, Capitol Reef National Park, UT (S. Dolan, 2001).

a Washington State apple historian, the 1930s tested the most committed of growers:

> *The Thirties were anything but kind to the apple grower. He was plagued by low prices, difficulty in getting loans to raise and harvest a crop, high freight rates, the ravages of the codling moth, the Teamsters Union interfering with his right to truck his own fruit to market, the Food and Drug Administration questioning the amount of spray residue on his apples, and high taxes. All these problems made it difficult for the best orchardist to survive. (Bright 88: 87).*

CCC's Role in Orchard Removal

Many orchardists did not survive the Great Depression, and the 1930s intensified the rate at which orchards were abandoned across the country, a trend begun in the 1880s. The Dust Bowl Era created severe drought throughout the Midwest and the South, and inundated orchards with wind-blown sand, killing fruit trees or ruining their crops (Figure 3.38). Across the United States 200,000 farms were abandoned or destroyed during the 1930s, and those that remained were harvested by poorly paid workers (Figure 3.39). The Civilian Conservation Corps (CCC), an Emergency Conservation Work program of the New Deal Era, contributed to the eradication of severely deteriorated orchards during the Depression. Among the scope of the CCC's conservation work was the reclamation of impoverished and abandoned farmlands. The CCC pulled out neglected fruit trees and reforested former agricultural lands in order to kill pests and diseases, conserve soil, and allow for recreational uses of the land.

CCC's Role in Orchard Replanting

Figure 3.38: *(above, left)* Historic photograph of wind-blown sand inundating an Oklahoma orchard in 1936. Two hundred thousand American farms were destroyed or abandoned during the Dust Bowl Era (courtesy of Library of Congress Prints and Photographs Division, Farm Security Administration, Office of War Information).

Curiously, the CCC also served to create orchards during their period of service from 1933 to 1942. At a number of historic sites managed by the National Park Service, the CCC planted orchards in order to restore a site to its appearance in a historic period. At Morristown National Historical Park in New Jersey, the CCC rehabilitated a deteriorated 18th century orchard by planting approximately 100 trees, among those the trees that remained from the earlier period (several of the oldest trees survived until the 1990s). At San Antonio Mission National Historic Site in Texas, the CCC planted fruit trees to re-create the appearance of 19th-century mission orchards. These were among the first reconstructed or "commemorative" orchards in the national park system, and their relative authenticity as representations of earlier orchards can be evaluated. However, as some of the earlier examples of cultural landscape treatment in the United States, commemorative orchards are potentially significant in their own right.

Obstacles to Citrus Development

Figure 3.39: *(above, right)* Historic photograph of peach pickers in Muscella, GA in 1936. Pickers received 75c/day (by Dorothea Lange, courtesy of Library of Congress Prints and Photographs Division, Farm Security Administration, Office of War Information).

The years from 1880 to 1945 were an experimental period for the new citrus, and nut industries. Both citrus and nut growing were initially thwarted by similar problems—a lack of good commercial varieties, difficulty in grafting and transplanting young trees, and intense competition from excellent imported fruit. The Florida Agricultural Experiment Station published three bulletins in 1913 intended to provide the fundamentals of citrus growing for the first time. In an attempt to boost growing in the state, the bulletins focused on the growing of tangerines, mandarins, lemons, and grapefruit in various parts of Florida. Some reliable varieties and methods of grafting could be recommended by this time. Eureka was by far the most popular variety of lemon.

The success of Navel orange in California, a variety that did not grow well in Florida, meant that California's citrus industry was ahead of Florida by 1913. California growers were struggling to meet the particular challenges of the new industry and stay ahead of citrus development in the Gulf states. Some of the challenges included the need for close regulation of soil fertility through the use of synthetic fertilizers to achieve a balance between vegetative growth and fruit production, and the need for great care in fruit storage to avoid infection by blue or green mold, both caused by *Penicillium* fungus. By World War II, California and Florida had become the centers of commercial citrus growing in the United States.

Citrus Orchard Characteristics in Period

Citrus orchards, or groves as they were also called, were laid out similarly to other orchards, with trees headed low and spaced according to the size of the species (Figure 3.40). Large-fruited trees, such as Navel orange and grapefruit, were laid out at the widest spacing of 25 x 25 feet. Smaller-fruited citrus, such as tangerines, mandarins, and lemons, were laid out at the smaller spacing of 15 x 15 feet (Lowther 14: III, 1476) (Figures 3.41 and 3.42). After World War II, oranges became the most widely planted orchard fruit in the United States, accounting for 24 percent of the total acreage of fruits grown (Census of Agriculture 80: I, 51).

Nut Development

The experimental years in commercial nut growing before World War II led to an important commercial industry in the latter 20th century. For pecan and almond, growth in commercial importance was almost as dramatic as for the orange. At the turn of the 20th century, the only American-grown nut with commercial value was the pecan, a native plant closely related to hickory, which had been grown extensively in Texas and other southern states for several hundred years. By 1980 pecans and almonds were the third and fourth most planted commercial orchard fruits, ranking only after oranges and apples (Census of Agriculture 80: I, 51). For American growers, successful nut production meant learning how to conquer the difficult task of grafting varieties of almond, pecan, English walnut, and European hazelnut onto seedling rootstocks, and then discovering how and when to transplant the young trees into the orchard without damaging their sensitive taproot (USDA 96: 10).

Figure 3.40: *(below)* Historic panorama of citrus orchards in Winter Haven, FL, in 1919, showing extensive plantings of low-headed trees (Library of Congress Prints and Photographs Division).

Figure 3.41: Photograph of the orange orchard at John Muir National Historic Site showing low-headed trees with wide spacing, typical of Muir's time. The trees are somewhat stunted, however, due to a nutrient deficiency, CA (S. Dolan, 2004).

Figure 3.42: Photograph of the lemon orchard at John Muir National Historic Site with low-headed trees, more closely spaced than the site's orange trees, CA (S. Dolan, 2004).

Regionalization of Nut Growing

As the early 20th century progressed, skills and knowledge about nut orchards were enhanced through research. California became the center of almond and walnut growing, with production gradually superseding the supply of French-imported nuts. Texas became the center of pecan growing. The Pacific Northwest, and in particular Oregon state, became the center of European hazelnut growing. New commercial varieties of pecan were developed in the southern states, new varieties of almond and walnut were developed in California, and new varieties of hazelnut were created in the Pacific Northwest. Nonpareil became the most widely planted almond variety, and Hartley was the most important walnut.

Nut Orchard Characteristics in Period

Variety development led to a great advancement of the nut industry; prior to this time only poorly adapted European varieties were available for each species. The new varieties bloomed later, avoiding frosts, and bore nuts younger and more abundantly. Very large and long-lived pecan and walnut trees were planted at 50 x 50 feet spacing, and small, peach-like almond trees were planted more closely at 20 x 20 feet. Hazelnuts were planted out at 10-15 feet in rows, with 15 feet between rows. Nut trees were headed low, similar to citrus and other orchard fruits of the period. However, walnut trees grown for both lumber and nuts were an exception to the low headed tree form.

Walnut Cultivation for Lumber

When grown for lumber, the walnut tree was high-grafted with a very tall, unbranched trunk. In this case the rootstock of black walnut, either the Eastern (*Juglans nigra*) or Northern California (*Juglans hindsii*) species would provide the valuable trunk wood, and the scion of English walnut (*Juglans regia*) would bear the valuable nuts. To optimize the development of lumber, the rootstock would form the black tree trunk and be grafted at more than six feet in height to the silver-gray scion. Lower limbs would be removed from the trunk to minimize knots in the wood. In the 50 or more years needed to grow harvest-worthy lumber, the walnut tree would provide a return with valuable nuts. These double-duty trees are distinctive in appearance from low-headed walnut trees grown just for nut production in this period. Several early 20th century specimens can be seen at John Muir National Historic Site in California.

Olive Development

The emergence of commercial olive orchards in California during the 1870s was contemporaneous with the early development of the citrus and nut industries. Like Spanish seedling oranges, the history of Mission olives in the United States dates to 1775 when Franciscan monks established the missions of San Diego, Santa Clara, and San Jose, and planted orchards of Spanish fruits. Now referred to as a variety, the 'Mission' olive was spread from the first California missions by cuttings and seeds and became planted throughout the Franciscan mission system of the late 18th and early 19th centuries. Planted in widely spaced orchards or in formal allées within mission grounds, Mission olives were grown primarily for their abundant oil. Olives were also used as pickled table fruits and were valued as graceful shade trees. Like most fruit trees of the period, Mission olive trees bore their canopies high on tall, unbranched trunks. Because olive cuttings rooted easily, trees started from cuttings were ungrafted; but because a tap-rooted seedling was difficult to transplant, trees started from seed were grafted (Taylor 00:19).

After the vast influx of Euro-Americans during the California Gold Rush, early California agriculture evolved from cattle ranching to wheat farming and then to fruit and vegetable growing. By the 1870s many farmers and nurseryman were seeking to differentiate their horticultural crops, so they turned their attention to the old abandoned olive trees in mission grounds, still alive despite decades of neglect. Sensing an opportunity to compete with imported olive oil, farmers and nurseryman began taking cuttings from the many remaining mission trees and creating stock for the first commercial olive orchards. The 1870s saw the planting of the first commercial orchards in the Central Valley, the Bay Area, and in coastal southern California.

Introduction of Manzanillo and Sevillano Olive Varieties

As California agriculture grew in profitability and sophistication, some wealthy growers sought out their own European olive varieties that could improve upon the qualities of the Mission. Benjamin Redding was one such importer, acquiring the Picholine variety from France in 1872. However, his efforts were surpassed by two more significant imports in the 1870s when Federico Pohndorf imported the Manzanillo and Sevillano varieties from Spain. These two new varieties and the old Mission olive would become the most abundantly grown varieties in the United States for the next hundred years.

The Manzanillo, or "little apple," olive had larger fruit and ripened earlier than the Mission, and produced both excellent oil and good table fruits. It was also more adaptable to humid conditions than the Mission olive, which tended to have problems with fungal diseases and insect pests in damp coastal areas. By the late 1880s, the majority of commercial olive growers in California were planting orchards of the Mission or Manzanillo varieties for oil production. The Sevillano, or "Queen," olive became more popular in the early 1900s after the development of canning technology gave rise to a table olive industry. The very large-fruited Sevillano was better suited to table fruit production than oil, and in the early 1900s many Mission or Manzanillo orchards were grafted over with the Sevillano (Taylor 00: 66).

The search for new olive varieties continued, however, and between 1880 and 1960 was led by the USDA and the University of California, College of Agriculture. During this period more than 300 olive varieties were imported from France, Italy, Spain, Greece, Turkey, and North Africa into the United States. Academic horticulturists and consulting nurserymen tested the new varieties for the characteristics of high productivity, abundance of oil yield, and resistance to temperature fluctuations and pests.

Olive Orchard Characteristics in Period

Despite rigorous horticultural research, no variety was found to replace the popularity of the Mission, Manzanillo, or Sevillano. Manzanillo became the most heavily planted variety in the San Joaquin Valley, and Sevillano was most important in the Sacramento Valley. The Mission olive was planted ubiquitously in olive-growing areas throughout the state. While academic research yielded only varieties that would have less economic importance, the USDA and University of California had greater influence on olive cultivation. Similar to their influence on the scaffold form of other orchard trees in this period, the USDA favored the "low-heading" of olive trees to promote easier access to the canopy and more youthful age of fruit production. Consequently, the appearance of olive trees changed from the tall trunk form of the trees depicted in historic mission photographs of the late 1700s and early 1800s, to the low-headed, short-trunk tree scaffold typical of most orchards in the late 1800s and 1900s.

The USDA and University of California also had an influence on the spacing of olive orchards. Horticultural research found that olives could be grown using dryland farming techniques, such as those practiced by John Muir in the northern Bay Area in the late 19th century. With a nod to the centuries-old knowledge of the Franciscans, horticulturists found that the tap-rooted olive tree could produce good yields without irrigation, provided the orchard was planted with wide spacing. Each olive tree had the ability to reach sufficient groundwater if given enough space to search. Between 1880 and 1945, wide spacing of 30 x 30 feet was common spacing in the unirrigated olive orchard, and 20 x 20 feet spacing more common in irrigated orchards. John Muir's olive orchard survives at the John Muir National Historic Site as an excellent example of a late 19th century dryland orchard in California. The low-headed trees are widely spaced, allowing their survival in dry conditions at the crest of a hill on Mount Wanda (Figures 3.43 and 3.44).

Ornamental Use of Olive Trees

As both a productive and aesthetic tree, the olive has performed double-duty for food and ornament in America. Planted along the alignment of cloister walks in missions, olive trees were regarded by the Franciscans as the graceful adornments of barren lands. Their tolerance of poor, dry soils

Figure 3.43: Photograph of a late 1800s olive orchard crowning a hill on Mt. Wanda, with former Chief of Interpretation David Blackburn, John Muir National Historic Site, CA (S. Dolan, 2005).

Figure 3.44: Photograph of low-headed trunks of widely spaced olive trees in the late 1800s. Inset shows developing fruit, Mt. Wanda, John Muir National Historic Site, CA (S. Dolan, 2005).

and hot climates led to use of the olive as a shade tree and as windbreaks in ranchos following the secularization of the missions. By the early 1900s, olive trees had become a status symbol of wealthy Californians with villas or country estates. Olive orchards were planted where they could be viewed upon arrival at the villa to create a romantic image. An excellent example of an olive orchard designed as part of a landscape plan survives at Filoli, a National Trust for Historic Preservation property in Woodside, California.

Purchased by the wealthy industrialist William Bowers Bourn II in 1916 to be a country estate, the landscape was designed with an olive orchard by Bruce Porter to create an Italian aesthetic (Taylor 00: 27).

Ubiquity of Commercial Orchards

By World War II, all of the common species of orchard fruits grown in the United States today had established industries fueled by commercial orchards. These fruit trees were grown as large or full-size, standard trees. After World War II, semi-dwarf and dwarf fruit trees would become the signature of a new era of intensive orchard management. The Great Depression and the war years influenced the transformation of the fruit tree, though curiously some influence came from the domestic rather than the commercial level. A popular interest in fruit and vegetable gardening emerged in the backyards of homeowners, stimulated in part by the vagaries of the Great Depression and fears of food shortages during World War II. Popular gardening literature encouraged homeowners to grow their own, more flavorful and nourishing produce than could be obtained from the store. Contemporary fruit offerings of the grocer were attacked as bland tasting and lacking freshness. A solution proposed by books, magazines, and radio shows of the war years was to plant a wide range of fruit trees in the backyard that would bear fresh fruit from the earliest sweet and sour cherries in June, to the latest apples in December.

Home Gardeners' Use of Dwarf Fruit Trees

Urban lots of the early 1900s were inadequate for the planting of more than one standard tree, but a solution for the gardener with limited space was the dwarf tree. Semi-dwarf and dwarf fruit trees, available to the formal fruit gardener since the early 1700s, were once again in vogue. Gradually plant nurseries began to propagate popular apple, pear, peach, and cherry varieties as semi-dwarf or dwarf trees to supply the growing demand from home gardeners. Apple was dwarfed by grafting onto the old Paradise and Doucin rootstocks; pear was dwarfed by grafting onto quince; peach and nectarine by grafting onto plum; and cherry by grafting onto an old slightly dwarfing cherry variety Mahaleb (Kairns 42: 18). At just 10 feet apart, a range of dwarf trees could be planted in the average backyard and they were easier to spray, prune, and harvest than the typical, full-size or standard tree. Dwarf trees also bore fruit younger than standard trees, in the second or third year, rather than in the fifth or even twelfth year.

While the small size of the family lot catalyzed the demand for dwarf trees in backyards sooner than their appearance in commercial orchards, in the post-war era the many favorable characteristics of dwarf trees would win their popularity with commercial growers too. During the war years, however, most growers continued to regard dwarf fruit trees with the same skepticism that Andrew Jackson Downing Jr. had expressed almost 100 years earlier. The 1914 words of horticultural writer Frank Waugh, in the magazine *Woman's Home Companion*, typified growers' attitudes towards dwarf fruit trees during the early 20th century:

> *It must be not understood that it is cheaper to grow fruit in this way, or that the dwarf trees are to take the place of standard fruit trees in money-making enterprises. We are talking of them now only as first class entertainment; they do [well] as an almost essential feature in the design of a city fruit garden.*

As the modern era in fruit growing drew to a close with the end of World War II, agricultural experiment stations were poised to lead growers into another era. An intense period of government-funded horticultural research in England and the United States was about to influence a shrinking of the scale of commercial orchards and a discarding of the archetypal, large-scale orchard as an archaic form of horticulture.

Most NPS Orchards Belong to this Period

The majority of old fruit trees and orchards within the national park system belong to this period in the history of orchards and fruit development. For many parks the period corresponds to the time when the land was in agricultural use before the designation of the park, and the period is sufficiently recent that it captures extant fruit trees with the greatest longevity, such as apple, pear, orange, and cherry. As the period contained the greatest loss in both the number of orchards and fruit varieties in the country's history, orchards and fruit trees that remain from this period are potentially significant as archaic forms of orchard horticulture, or for the presence of rare fruit varieties or strains of varieties that since have been lost from American horticulture.

Summary

The evolution of orchards from 1881 to 1945 was fueled by technological and scientific discovery, and led to the professional and commercial development of the orchard industry. The most important changes from a cultural resource management perspective were transformations in the form, shape, and layout of orchard trees, and a dramatic reduction in the number of varieties grown. These transformations were influenced by the involvement of the federal government in horticultural development through the creation of the United States Department of Agriculture and the establishment of agricultural experiment stations in every state.

Orchard tree form changed from a five-foot-tall trunk to a less than three-foot-tall trunk; tree shape was changed from an unpruned, natural state to either a central leader or open-bowl pruning style, and orchard layout was expanded to greater spacing. The layout of apple and pear orchards varied from 30 x 30 feet spacing to as much as 40-50 x 40-50 feet, and for tighter-spaced fruits such as peach, plum, and cherry, the layout was changed from a square to a rectangular arrangement. The layout changes were made for greater access for new machinery and equipment and to increase the yield from mature trees.

The dramatic decrease in the number of varieties grown was due to a process of selection for commercial fitness. Commercial criteria were refined during the period to incorporate all factors promoting high yields and durability of harvested fruit. As a result the number of varieties of all orchard fruits grown was pared from many hundreds to tens. At the end of the period, most orchard fruit species were represented by just 10 widely grown commercial varieties. For most fruit species, the top 10 varieties were dominated by one variety with the greatest commercial value and most widespread planting. For apple, Baldwin and Ben Davis were the most important commercial varieties in the early 20th century, but were rapidly superseded by McIntosh for Baldwin and Red Delicious for Ben Davis.

The development of Red Delicious during this period had an enormous impact on apple growing, resulting in greater profitability for the industry, great fashionability of red apples, greater ubiquity of a single variety, and further obsolescence of superseded varieties. The Bartlett variety dominated the pear industry, with New York and California as the greatest producers. Meanwhile, an expanding industry in the Pacific Northwest favored Anjou and Bosc varieties.

Peach growing was ubiquitous with many local varieties remaining important, though Elberta became the most dominant and widely grown variety. For cherry and plum, commercial growing became regionalized rather like pear. Sour cherry production became centered in the upper Midwest, where Montmorency was the dominant variety, and sweet cherry production was taken over by the Pacific Northwest where Bing was the most important variety. Plum growing was not dominated by a single variety, but specialization occurred in certain regions. Japanese and European plum growing centered on the West Coast, American varieties were grown in the Midwest and the South, and European varieties were grown to a less extent in the eastern states.

Citrus, olive, and nut species were the only orchard fruits to have a net increase in their number of varieties during this period. These industries were born and established during the period as a result of the development or importation of varieties and breakthroughs in the horticultural techniques of propagation and transplanting. The citrus, olive, and nut industries also became highly regionalized with citrus centered in Florida and California and olive production in California. With commercial nut orchards, pecan production was centered in Texas, almonds and English walnuts in California, and European hazelnuts in the Pacific Northwest.

The number of fruit trees and orchards fell dramatically during the period, with all but the West Coast losing orchards to increasing urbanization. Approximately 50 percent of the fruit trees that existed in 1880 were gone by 1930, though the great paring down in the number of orchards was paralleled by a rise in specialized, commercial orchards, managed by growers rather than farmers. Technologies that buoyed the development of commercial orchards included a nationwide network of railroads and then later a national system of roads, growth in canning technology and irrigation systems, and the discovery of mechanical refrigeration and cold storage. Scientific breakthroughs included the discovery of disease organisms and the development of the first pesticides for orchard pest control.

Orchard management became more elaborate with the standard practices of pruning, spraying, fertilizing, and cultivation. Packing sheds were added to western orchards, and packing containers and advertising were designed to increase profitability. The Progressive Era brought regulatory challenges for growers and the Great Depression brought economic hardship, both resulting in greater cooperation between growers through the development of growers' cooperatives and marketing commissions. Packing sheds were lost from western orchards as centralized fruit handling facilities were created, and still more orchards were abandoned during the Depression years when almost no growers turned a profit. The Civilian Conservation Corps contributed to the removal of orchards between 1933 and 1941 through reclamation efforts, but also served to create orchards by replanting commemorative orchards at national historic sites.

At the end of the period, the apple variety Golden Delicious was steadily gaining popularity with growers as a potentially lucrative new yellow apple and a good pollenizer for the dominant, self-sterile Red Delicious variety. Dwarf fruit trees were becoming popular with urban and suburban homeowners with fruit and vegetable gardens, and intense research was underway to move the industry towards smaller trees and intensive orchard management. Full-size, standard trees remained the norm for all commercial orchard fruits, although the need for labor savings through more accessible orchard trees was becoming very evident.

CHAPTER 4: FRUIT MONOCULTURE AND ORCHARD INTENSIFICATION: 1946-PRESENT

Internationalism, dwarf trees, and high-density management systems.

Purpose of Contemporary Context

This chapter explores how distinctly different contemporary orchards are compared with orchards from earlier periods. By identifying these differences it's easier to understand how orchards from earlier periods may have historic significance, and be eligible for listing in the National Register of Historic Places. As described in Chapter 5, orchards that date from the earlier periods may have historic significance and physical integrity, and may be eligible for listing in the National Register. Whereas orchards that date from the present period generally embody the current characteristics of orchard design, and are therefore unlikely to be historically significant.

Characteristics of Change

The current period of fruit monoculture and orchard intensification began at the end of World War II. The end of hostilities saw a new generation of growers emerge and, for the first time in 60 years, an increase in the number of American orchards. Through the efforts of this educated generation of growers and their teachers from a national and international network of academic institutions, orchards would evolve more drastically in appearance than in any former period. The overall change can be generalized as a shrinking of the size of orchard trees, a great reduction in the spacing between trees and between rows, and a vast increase in the number of trees per acre.

Concurrent with intensification was orchard monoculture, the mass planting of large blocks of singular commercial varieties for hundreds and sometimes thousands of acres. This trend reached its zenith in the 1980s when consumers began to demand greater choice in the number of varieties. Curiously, the recent trend towards more varieties was fostered by the use of dwarf fruit trees and consumer demand for more choice. This period also saw a big increase in the number of fruit species grown in the Mediterranean and tropical and subtropical climates of the United States, including fruits outside the scope of this study, such as guava, avocado, pistachio, and lychee fruit.

End of Farm Orchards

The current period in the history of orchards in the United States saw the end of farm orchards as amateur, small-scale commercial and subsistence enterprises. During this period new orchards were planted as professional commercial orchards or as small backyard fruit gardens, not as small parcels of mixed fruit species and varieties of one to five acres in size. This change accompanied the nation's shift from an agricultural to a mixed industrial and technology-based economy, and from rural to urban, and then suburban lifestyles. In the late 1900s, a highly competitive economic environment for commercial orchards created very small profit margins for orchardists; this drove the physical changes to orchards. During this period commercial orchards were highly pressured business enterprises commanding great economic investment and large risks to turn a profit. With such constraints, orchardists were compelled to dedicate themselves to business plans as well as to the state of the art in orchard culture in order to keep pace with extremely competitive market forces and maintain economic viability.

High Costs of Commercial Production

With the coming of peacetime after World War II, increased consumer confidence and buying power stimulated greater demand for quality fresh produce. New orchardists who perceived a potentially lucrative opportunity soon encountered an economic reality that emerged soon after the turn of the 20th century: costs of production were high and rising. Despite great demand for fresh fruit, a growing number of competitors and high production costs could reduce orchardists' profits to the point of bankruptcy. In the contemporary orchard, the costs of pruning, fertilizing, spraying and mowing, now perceived as fundamentals in orchard management, were rising sharply along with wages, energy prices, and the cost of manufactured fertilizers and pesticides. Growers recognized that many of these routine maintenance practices were similar in cost per acre regardless of the size of the crop, but as per-acre cost of production increased, the per-bushel profit decreased. The perceived solution to high costs was to squeeze more bushels out of the same acre of land.

Impetus to Dwarf

In order to increase yields, researchers and practitioners turned their attention to modifying the unit of production itself: the orchard tree. By the 1950s it was recognized that major advancements had occurred in orchard technology through innovations in equipment, pesticides, growth-regulating chemicals, fruit-handling techniques, packing systems, and storage. However, little change had been made to the actual production unit, the tree itself (Carlson 70:181). The major innovation in tree form in the former 60 years was the practice of low-heading or shortening of the tree trunk, from 5-8 feet down to 18-36 inches. This lowered the tree canopy for easier access and stimulated the tree to fruit within five to seven years after planting, rather than 10 to 12 years with a tall trunk tree on a seedling rootstock (Tukey 64: 85). However, it was widely acknowledged that low-headed trees were still highly laborious to work with, and consensus was emerging among horticulturists that big trees did not equate with big efficiency. The large canopies of low-headed fruit trees prevented optimal light interception by the total leaf area, and shading in the lower part of the canopy resulted in poorer quality fruit (Carlson 70: 176). It was thought the best way to tackle the problem of yield and profitability was to shrink the size of the tree.

Transition to Dwarfs

Between 1945 and the present, almost all species of orchard fruits were transformed from full-size or standard trees to dwarfed trees grown at closer spacing. This was done through the development or standardization of dwarfing rootstocks, through the use of chemical growth regulators, or through the use of spur-type varieties for some apples and pears. Spur types are the mutant strains of some common apple and pear scion varieties that naturally form semi-standard sized trees by having shorter internodes and more fruiting spurs. Spur-type varieties will be discussed further in this chapter, but it is important to emphasize that the use of spur types represents one of two tracts of transition from standard to dwarf trees. The tract with spur types involves the scion and is distinct from the tract involving dwarfing rootstocks. The use of spur types with apple and pear was most important from the 1950s to the 1980s, whereas the use of dwarfing rootstocks physically evolved in American orchards from the 1950s until the present. The adoption of dwarfing rootstocks has taken the last 50 years to encompass nearly all fruit species, such as apple, pear, plum, apricot, cherry, nuts, and citrus. However, for a few varieties of apple and pear, the use of spur types was a second, more short-lived but concurrent transition.

Need for Standardization

Ultimately, the use of clonal, or cloned, dwarfing rootstocks became the industry standard for the culture of almost all orchard fruit trees. Only peach is currently lacking a clonal dwarfing rootstock. However, it has been adapted to the same intensive system of orchard management as other fruit species, using standard peach trees with very close spacing and heavy summer pruning (Barritt 01: interview). Before dwarfing rootstocks could be made available through plant nurseries, standardization was first necessary. Standardization involved selecting rootstocks with a particular performance in dwarfing the scion, then classifying and cloning them through vegetative propagation. Up until the 20th century, no standardization existed for identifying dwarfing rootstocks. Dwarfing rootstocks had existed for as long as fruit tree culture, but their availability to growers and performance was unreliable. Standardization of rootstocks began in the 1910s in Germany, the Netherlands, France, and England, with the work in England having the most significant influence on American orchards in the later 20th century (Tukey 64: 128).

Standardizing Dwarfing Rootstocks

In 1912 the East Malling Horticultural Research Station in the Wye Valley of south England began collecting cuttings of dwarf seedling apple trees from all over Europe. Cuttings were rooted, grafted as rootstocks to a range of scions, and their influence on the growth characteristics of the scion varieties observed over a period of years. Between 1912 and 1917, Sir Ronald Hatton, the lead horticulturist, oversaw a team of researchers at East Malling to trial hundreds of apple trees and select just 16 types as dwarfing rootstocks in the early 1920s (Carlson 70: 161). The first 16 classified with the Roman numerals EM. I to EM. XVI were selected for their individually unique abilities to influence the vigor or dwarfness of a scion (Figure 4.1). The majority of the 16 dwarfing rootstocks were variants or strains of the Paradise and Doucin apple varieties that were known in Europe in the 15th century (Tukey 64:13). Another 10, clonal dwarfing apple rootstocks, EM. XVII to EM. 26, were created at East Malling between the 1920s and 1950s, by which time the naming convention was changed to Arabic numerals to avoid the complexity of the higher Roman numerals (Tukey 64: 133).

Figure 4.1: Historic photograph of Sir Ronald Hatton, Director of East Malling Horticultural Research Station, England, with semi-dwarf Bramley apple trees grafted onto M. 7 clonal rootstocks, in 1935. The trees were 15 years old when photographed (from Tukey, 1964, courtesy of Cornell University Press).

Introduction of Apple Dwarfing Rootstocks

The East Malling clonal dwarfing apple rootstocks were first imported into the United States in 1929, just before a federal government embargo would have prevented their admission. In 1930, a federal embargo prohibiting the importation of orchard rootstocks went into effect to prevent the spread of pests and diseases through the country's agricultural crops. The Agricultural Experiment Stations in New York, Massachusetts, Pennsylvania, and Ontario, Canada collaborated to obtain the first 16 East Malling apple rootstocks before the ban was imposed. Trials were then conducted to select the best rootstocks for North American apple growing conditions, and those selected were propagated for dissemination. In 1937 the first dwarf apple trees on East Malling clonal rootstocks were available for sale by some nurseries in the United States (Tukey 64: 26).

While horticulturists had their first opportunity to test the East Malling clonal rootstocks on American soil, work continued in England to expand and improve the range of dwarfing rootstocks. Greater constraints on the size of agricultural land parcels and the premium nature of good orchard sites led to an earlier interest and commitment to using dwarf fruit trees in Europe than in the United States. Before World War II, growers throughout Europe, Australia, New Zealand, and parts of Asia began the conversion from standard apple trees to dwarfs.

Rootstocks and Disease Resistance

In 1992, the East Malling Research Station began to collaborate with the John Innes Horticultural Institute of Merton to develop a new series of clonal dwarfing rootstocks for apple that conferred disease resistance to the scion. The new research project demanded a greater level of sophistication than the first, involving systematic breeding or hybridization of parent apple trees to create offspring that conferred dwarfing and disease resistance. Resistance to woolly aphid, *Eriosoma lanigerum* was one of the primary objectives of the project, as this insect was an economically significant pest of apple orchards in Australia. The American apple variety, Northern Spy, was used as one parent in the hybridization work since it was known to be resistant to aphids. The work of the two research stations intensified in 1928 and by 1952 the first Malling-Merton or "MM" rootstocks were available to growers. Out of 3,758 hybrid progeny, only 15 MM rootstocks were selected, numbered from MM. 101 to MM. 115. Of these, only MM. 106 and MM. 111 would have economic importance in the United States in later 20th century apple orchards. In addition to apple rootstocks, the Malling and Merton Research Stations began investigating dwarfing rootstocks for pear, cherry, peach, nectarine, apricot, plum, and citrus before World War II.

East Malling clonal rootstocks did not transform the appearance of American apple orchards overnight. Although dwarfing rootstocks had become more widely available by the late 1950s and early 1960s, dwarf trees demanded specialized knowledge, skills, and equipment for their effective use. Many growers held on to their standard trees, letting others take the plunge and make the mistakes associated with use of new technology. Growers understood, for example, that not all varieties could be grown on all EM or MM rootstocks due to "scion incompatibility," and not all rootstocks were suitable for all growing conditions. Some, for example, could not tolerate light soils and droughty conditions, while others created such miniaturized, fragile trees that the trunk needed staking for the life of the tree or it would collapse under the weight of fruit or develop wind-throw (Figure 4.2). Dwarf trees were thought of as meticulous "garden trees" from their popularity with homeowners, and would not tolerate any neglect in a commercial orchard.

Characteristics of Dwarf Trees

Eastern growers were the first to receive and use the East Malling rootstocks, and the strength of their industry relative to the Midwest and the West helped them make the earliest transition to dwarfing rootstocks. By the early 1960s, eastern apple growers were using some of the East Malling clonal rootstocks, and EM. VII, later known as M. 7, and MM. 106 were the most popular. These rootstocks created a semi-dwarf tree, with M. 7 producing a tree 50 percent of the standard size, and MM. 106 slightly larger (Figure 4.3). Six times as many trees per acre could be planted than with standard trees, at 10 x 18 spacing and 240 trees per acre. In contrast, standard trees required 30 x 30 feet spacing or 40 trees per acre. M. 7 and MM. 106 had broad compatibility with a wide range of apple scion

Figure 4.2: Photograph of a dwarf McIntosh apple orchard grafted to clonal dwarfing rootstocks, leaning due to wind-throw from lack of staking (from Tukey, 1964, courtesy Cornell University Press).

Figure 4.3: Photograph of low-headed semi-dwarf apple trees at Capitol Reef National Park, UT. Trees were grafted onto clonal dwarfing rootstocks, such as M. 7, which limited tree size and miniaturized stature. Trees were 15 years old when photographed. Semi-dwarf trees are incompatible with the period of significance of the park's Fruita Historic District (S. Dolan 2001).

Figure 4.4: *(facing page, left)* Publicity photograph of a 3-year-old dwarf Baldwin apple tree, demonstrating youthful age of fruit bearing. The tree and girl were the same age, but the tree "would never grow taller than a man could reach." Here, fruit-laden branches touch the ground as the trunk is too short (from Tukey, 1964, courtesy of Jackson and Perkins Company).

Figure 4.5: *(facing page, right)* Photograph of a semi-dwarf Red Delicious apple tree with M. 7 or MM. 106 rootstock and a relatively taller trunk to prevent fruit-laden branches from touching the ground. This modern period, semi-dwarf tree is incompatible with the character of fruit trees in the period of significance of the Adams orchard. Heidi Cope, Horticulturist, provides scale, Adams National Historical Park, MA (S. Dolan, 2001).

varieties and soil conditions, and required no staking. The only perceived disadvantage of M. 7 was a tendency for the rootstock to sucker. Vertical water sprouts borne from the base of the tree would crowd and compete with the scion. This problem could be overcome by using trees with a higher graft union between the rootstock and the scion. In this case, the nursery grafted the scion at 16 inches high on the rootstock whip (single, unbranched stem), so the rootstock could be buried deeper at planting, discouraging the roots to sprout suckers.

The size of the tree and height of the graft union weren't the only different aspects of the form of the clonal dwarf tree. The optimal form of a tree on a clonal dwarfing rootstock was found to be the "central leader" or "modified central leader" rather than the "open bowl," as found in trees on seedling or standard rootstocks before World War II. The tendency of the dwarf tree to bear fruit early caused the lower branches to sag from the weight of the fruit. To prevent heavy branches from touching the ground, it was necessary to form the lateral branches proportionately higher on the trunk (Teskey 78: 185) (Figures 4.4 and 4.5). After more than 60 years of short trunks, the trunks of fruit trees were becoming proportionately taller.

Spread of Apple Dwarfing Rootstocks

In 1958 dwarf fruit tree researchers and growers in Michigan formed the International Dwarf Fruit Tree Association (IDFTA), marking the beginning of international influence on dwarf fruit tree development and use in the United States. A network of academic institutions and growers, the organization served as a more informal conduit for sharing international dwarf fruit tree research with American orchardists than through traditional government-to-government liaisons. The IDFTA held annual conferences and published quarterly journals titled *The Compact Fruit Tree*. During the 1960s, the use of clonal dwarfing apple rootstocks spread throughout the United States, reaching the West by the late 1960s. In the Pacific Northwest, the National Park Service (NPS) used apple trees grafted to clonal dwarfing rootstocks for a reconstruction orchard at Fort Vancouver National Historic Site. The orchard, depicting the early 19th century Hudson's Bay Company's fort orchard, was funded by the Mission 66 initiative.

Influence of NPS Mission 66

Spanning from 1956 to 1966, Mission 66 was a huge capital improvements campaign for national parks approved by Congress to counter deteriorating park conditions after World War II. Justified in part by rapidly increasing park visitation and billed as a celebration of the 50th anniversary of the NPS, the Mission 66 program enabled several historical parks to reconstruct orchards that had vanished. However, because the Mission 66 program preceded the National Historic Preservation Act—and its concept of historic integrity and professional preservation standards—these early reconstruction projects didn't necessarily result in orchards that were authentic to the historical period they were meant to depict.

At Fort Vancouver, for example, the Mission 66 orchard planting used 20th century apple varieties grafted onto the semi-dwarf M. 7 rootstock rather than the 19th century varieties on seedling rootstocks that were used by the Hudson's Bay Company from 1825 to 1846. While the trees used in the reconstruction were typical 1960s' Pacific Northwest nursery stock, the semi-dwarf trees did not accurately depict the landscape's period of significance. They were too short, too tightly spaced, and of varieties that didn't exist in 1846. Another example of Mission 66 orchard reconstruction inaccuracies can be seen at the Hopewell Furnace National Historical Park in New Jersey. Here an orchard was planted to reconstruct a 19th century farm orchard, and while the planting accurately used 19th century apple varieties on seedling rootstocks, the trees were planted in the context of a new visitor center parking area to serve both interpretive and landscaping purposes. The accurately high-headed or tall-trunk trees were implausibly laid out on the slopes of the parking area's graded berms, rather than on the level orchard floor that existed historically. Other examples of Mission 66 "historical" orchards can be found throughout the national park system, and they often present a modern interpretation of period orchards that was unconstrained by the concept of historical integrity. These orchards reflect the history of historic preservation in the NPS, and may become historically significant in their own right.

Influence of Harold Tukey

In 1964 Harold B. Tukey, former President of the American Society for Horticultural Science, Professor of Horticulture at Cornell University and a founding member of IDFTA, authored the seminal work *Dwarfed Fruit Trees for Orchard, Garden, and Home.* In this work Tukey laid out specific directions for the use of dwarf trees and clearly identified their future in the commercial industry. His promotion of their benefits for all fruit species, not just apple, helped to buoy the new wave of interest in smaller trees. Tukey's book highlighted a long list of benefits of dwarf fruit trees over standards. His views were derived from four decades of research at the New York Agricultural Experiment Station and from the work of his colleagues, nationally and internationally. So compelling were his arguments in favor of dwarf trees that few horticulturists posed a dissenting viewpoint.

Advantages of Dwarf Trees

One of the great benefits of dwarf trees, Tukey asserted, was to provide regular and annual production, rather than the somewhat irregular and often biennial production of standard trees. Dwarf trees provided an early return on the grower's investment due to their earlier age of fruiting and the greater number of trees per acre. Benefits also included lower costs per unit of production and reduced labor hours due to easier and more economical management of the orchard (e.g., in pruning, fruit thinning, and harvesting) (Figure 4.6).

Tukey indicated that pest and disease control would also be cheaper and more efficient with dwarf fruit trees due to ease of spraying and less volume of spray needed. In addition, clonal dwarfing rootstocks could be guaranteed virus free by the late 1960s (called "EMLA" rootstocks). Dwarf trees also produced a higher percentage of high-quality fruit with fewer culls. This was due to increased light interception by the foliage and fruit, improving fruit color and greater uniformity in fruit size. The smaller tree scaffold also reduced fruit blemishes and bruising due to less woody tissue to abrade the fruit and better pest control (Tukey 64: 32-48).

Variety Obsolescence

Harold Tukey predicted the 21st century future of the American commercial orchard industry when, in 1964, he stated that orchardists should adjust to changing conditions and market demands. Presciently, he demanded action to prevent fruit variety obsolescence, which would occur through consumers tiring of only a few varieties on the market. Dwarf trees, he stated, could provide this flexibility by being more easily replaced with new varieties. While the productive lifespan of a standard apple orchard was

Figure 4.6: Photograph of an orchardist demonstrating the ease of harvesting a 5-year-old dwarf McIntosh apple tree, grafted to a M. 9 clonal dwarfing rootstock (from Tukey 1964, courtesy of Cornell University Press).

thought to be 25 to 40 years by the 1960s, Tukey argued that dwarf trees made it possible to "rotate" an orchard every 15 years or less, as they were younger to bear fruit and become profitable. While unstated by Tukey, the shorter lifespan of dwarf trees actually necessitated their earlier removal than standard trees. But practically speaking, dwarf trees could be more easily torn out and replanted with a new variety to keep pace with economic trends (Tukey 64: 43). While his promotion of the benefits of dwarf trees had great influence upon growers, his ideas about flexibility with market conditions were unheeded until the 1980s. Only after American orchardists were backed into an economic corner through reliance on a handful of fruit varieties were Tukey's insights heeded (Phillips 04: 1).

Mid-1900s Dominant Apple Varieties

The popularity of a handful of varieties with growers was due in part to the successful commercial characteristics of those varieties. For example, by the 1960s the top 10 apple varieties grown in the United States were all winter varieties with the exception of Jonathan (a fall apple), as winter varieties had better keeping qualities and performed well in either cold temperature or the new Controlled Atmosphere (CA) storage conditions. In descending order of importance the top 10 varieties were: Red Delicious, Golden Delicious, McIntosh, Jonathan, Rome Beauty, Winesap, Baldwin, Northern Spy, Rhode Island Greening, and Yellow Newtown Pippin. Almost one-third of all apple trees grown were Red Delicious, with Golden Delicious a close second (Carlson 70: 33). The popularity of the dominant apple varieties with growers was also due partly to the existence of spur types. Red Delicious, Golden Delicious, McIntosh, and Winesap, among others, had spur types and this contributed to their favor with growers and the slower adoption of dwarfing rootstocks in the United States than in Europe.

Advent of Spur Types

Between the 1950s and 1970s, spur-type strains were created in a handful of the most popular apple varieties and to a smaller extent with pear varieties, leading to this period being named the "Red Sport, Spur Type Era" by fruit historian Virginia Maas. During this period, red sports of apple and pear varieties with greater red coloration were popular with consumers, and heavy-cropping spur types were popular with growers (Carlson 70: 63). The first spur-type strain, a naturally occurring mutation of a variety, was discovered by an orchardist in eastern Washington in 1948. The grower noticed that one Red Delicious tree in his orchard was more heavily covered with fruit than its neighbors, as the tree had more fruiting spurs or short lateral branches producing fruit. The abundance of spurs was caused through a shorter internode length, the distance between axillary buds. The shorter internode length resulted in a tree that was naturally dwarf, or actually semi-standard, according to Tukey's classification of dwarf tree sizes, at two-thirds the size of a standard Red Delicious tree (Figures 4.7 and 4.8). Tukey used the descending scale of standard, semi-standard, semi-dwarf, and dwarf to identify the range of fruit tree sizes.

Characteristics of Spur-Type Orchard

During the 1950s, many more spur type mutations of Red Delicious were discovered, with the Starkrimson Red Delicious (patented by Stark Brothers Nursery) being one of the best known. Starkrimson could be planted at 10 feet apart in rows 20 feet apart, slightly wider spacing than with M. 7 or MM. 106 semi-dwarf trees, and yield 400 bushels or 16,800 pounds of fruit per acre, a great improvement upon 250 bushels per acre with standard trees. The first spur type of Golden Delicious, Starkspur, was discovered in 1955 (also in eastern Washington) and the first spur type of McIntosh—Mac Spur—was discovered in 1969 in British Columbia (Carlson 70: 106). While some new spur-type varieties of pears were selected in this period, such as the 1960-created European pear variety, Moonglow, and the Asian pear variety Niitaka, the most important pear varieties were still non-spur types.

Figure 4.7: *(below, left)* Photograph of a spur-type Golden Delicious apple tree branch, showing heavier fruit bearing during to an abundance of fruiting spurs (from Tukey, 1964, courtesy of Cornell University Press).

Figure 4.8: *(below)* Photograph of a standard Golden Delicious apple tree branch, showing lighter fruit bearing by comparison to the spur type tree (from Tukey, 1964, courtesy of Cornell State University).

Influence of Spur Types

Curiously, Red Delicious proved to be the most highly variable of all apple varieties, bearing the most frequent mutations and yielding more than 400 red sport and spur-type strains between the 1920s and 1980s. The influence of spur-type strains was for growers to select the naturally semi-standard spur-type strains of Red Delicious, Golden Delicious, and McIntosh, for example, over other non spur-type varieties that would need to be grafted to clonal dwarfing rootstocks to reduce their size. Borne on seedling rootstocks, spur-type scion strains would produce more fruit than standard trees, and could be grown without the specialized knowledge, skills, and equipment of trees on dwarfing rootstocks. With spur types, growers could perpetuate the same pruning practices they had used for standard trees, that being low-heading in the open bowl or central leader shape. They could also avoid investing in new equipment, such as more compact tractors and smaller spray rigs that were required for the tighter spacing of dwarf trees. Spur-type strains provided a less demanding alternative to American apple and pear growers than trees on clonal dwarfing rootstocks. As a result, spur types influenced a slower transition to dwarf trees than in Europe (Barritt 01: interview).

Advent of Red Sports

The emphasis placed on discovering spur types of the popular varieties was perhaps surpassed only by the excitement generated by the selection of red sports. Between the 1920s and 1980s, the American nursery trade generated greater revenues by discovering and patenting naturally occurring red strains of varieties. From the first red sport of (Red) Delicious, Starking, in 1921, to the selection of Jonnee, a red sport of Jonathan in 1967, almost every popular apple and pear variety had red sports. Among the best known are Red Spy of Northern Spy, Red York Imperial of York Imperial, Red Rome of Rome Beauty, Red Bartlett of Bartlett, and Red Anjou of Anjou.

Red sports are mutations with a heavier solid blush on their fruits than their counterparts, and will color sooner and can be picked earlier on the tree. Their skin has the appearance of greater ripeness than the stage of development of their flesh, and earlier picking allowed for more control of the ripening process in CA storage. Red Delicious' hundreds of red sports contributed to its unprecedented rise in popularity. The variety was considered America's most attractive red apple and most valuable apple export. However, the increasingly thick black-red skin and ever-whiter flesh of the many red sports, and then of the red sports derived from red sports, eventually contributed to its fall from favor. W. A. Luce, former President of the American Pomological Society, foreshadowed the demise of Red Delicious, when he stated in 1969:

> *Many growers believe that the Standard Delicious is a better flavored apple than the new redder strains; one reason perhaps might be that the Delicious was generally left to hang on the tree longer to get color and gain more maturity and flavor.* (Carlson 70: 54).

Luce's words are curiously contrasted with a statement by Luther Burbank, California's famous horticulturist, made 40 years earlier:

> *[Red] Delicious is the best in quality of any apple which I have so far tested.* (Carlson 70: 50).

Creation of Pear Dwarfing Rootstocks

The intensification of American pear orchards evolved more slowly than apple orchards. The adoption of dwarfing rootstocks for commercial pear orchards was primarily a western United States phenomenon, reflecting the relocation of the industry to the West from 1945 to present. Use of dwarfing rootstocks for commercial pears began before World War II, but did not become a strong trend until the 1980s. As described in Chapter 3, fire blight epidemics virtually eradicated the commercial pear industry in the eastern United States in the early 1900s, and by the 1970s, 90 percent of the industry was located west of the Rocky Mountains. In the late 1900s the industry was split between California and the Pacific Northwest, with 50 percent of all commercial pear orchards located in California, 30 percent in Washington, and 20 percent in Oregon (Teskey 78: 127). Fire blight proved less pernicious on the West Coast due to drier conditions during blossoming in California, or to cooler temperatures at blossom time in Washington and Oregon.

Before World War II, California growers began to attempt to control the size of the naturally large pear tree, Comice, by grafting the variety onto quince rootstock. Eastern United States and European pear growers had known about the ability of quince, *Cydonia oblongata,* to dwarf pear for several hundred years. However eastern growers had limited economic success with quince as a pear rootstock due to its lack of cold hardiness and susceptibility to fire blight. The East Malling Research Station began to study quince rootstocks in 1914, and made available the first standardized clonal quince rootstocks for pear in 1920. Of the seven types originally selected, only three, EM. Quince A, B, and C, would have commercial importance, and only two, EM. Quince A and EM. Quince C, would be significant in the United States.

The East Malling quince rootstocks were selected from naturally occurring varieties of quince that grew in France, named the Angers, Provence, or Fontenay varieties for their area of origin. EM. Quince A (Angers) was the

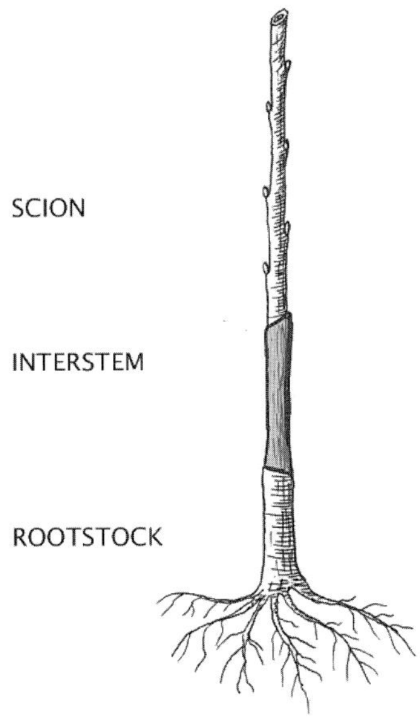

Figure 4.9: Drawing of a young tree with an interstem grafted between the scion and rootstock, to prevent incompatibility between the scion and rootstock (from Tukey, 1964, courtesy of Cornell University Press).

least dwarfing, creating a pear tree about half the size of a standard. EM. Quince B (Provence) created a pear tree slightly smaller than Quince A, and EM. Quince C (Fontenay) produced a tree about one-third the size of a standard pear. EM. Quince A, B, and C produced pear trees that were not self-supporting and had to be staked, with the exception of Comice, a particularly robust variety that could support itself on quince roots (Teskey 78: 174).

Like apple orchards, the standardization of dwarfing quince rootstocks for pear did not change the appearance of American pear orchards overnight. Challenges existed in the use of the quince rootstocks that required adaptability by nurserymen and growers. The challenges were due to quince's susceptibility to fire blight, its incompatibility with various scion varieties and its need for staking. EM. Quince A was found to be incompatible with the Bosc, Bartlett, and Seckel varieties, and EM. Quince C was incompatible with even more because it did not develop a strong graft union with the scion. Horticulturists discovered that Bartlett, Bosc, and Seckel could be grown on EM. Quince A if an intermediate stem, or interstem of the varieties Beurre Hardy or Old Home, was grafted between the scion and the quince rootstock. The interstem, a six- to eight-inch stem of the compatible variety, would serve as a bridge between the scion and the rootstock, allowing both to be joined (Figure 4.9). Interstems of Beurre Hardy and Old Home were also found to confer fire blight resistance to the quince rootstock.

Slow Adoption of Pear Dwarfing Rootstocks

Despite these horticultural discoveries in the development of dwarfing rootstocks for pear, the majority of pears from the 1950s to the 1970s were grown on standard rootstocks and not clonal quince rootstocks. Standard rootstocks were derived from the seedlings of the European pear *Pyrus communis*, with the seeds commonly obtained from Bartlett pear processing plants or canneries. The widespread popularity of the Bartlett variety, as a naturally smaller-sized scion than other pear varieties, influenced the slower adoption of dwarfing rootstocks for pear in the United States. By the 1960s, 90 percent of commercial pear trees in France were grafted to dwarfing quince rootstocks, whereas less than five percent of American pear trees were dwarfs. Standard pear orchards were grown with a spacing of 25 x 25 feet, and dwarf pears on quince rootstocks could be grown at 10 x 12 feet for EM. Quince A and 8 x 14 feet (a hedgerow form) for EM. Quince C.

Full Adoption of Apple Dwarfing Rootstocks

By the late 1970s, the transition from seedling apple rootstocks to clonal dwarfing rootstocks was ubiquitous. Semi-dwarf M. 7 and MM. 106 clonal rootstocks were widely used in new apple orchard plantings, and the abandonment of spur type varieties had begun (Barritt 2001: interview). The transition was driven by a newly introduced USDA concept of "rootstock efficiency," an expression of the optimization of land for fruit production. The concept, defined as the number of fruits produced per cross sectional area of trunk, demonstrated that clonal dwarfing rootstocks, rather than spur type varieties or seedling rootstocks, produced more fruit per unit of wood in each trunk. Varieties grafted onto clonal dwarfing rootstocks would produce more fruit per unit of wood production, whereas spur types and seedling rootstocks produced relatively more wood per unit of fruit.

Development of the concept of rootstock efficiency represents the increasing sophistication in orchard horticulture at this time in response to market conditions. The new concept examined the fruit yield to wood quantity at different ages after planting, rather than the fruit yield to land acre relationship that was of concern before World War II. In the new period, fruit yield per acre had become relatively meaningless. Only quality fruit yield counted for the fresh market, and the greatest amount of quality fruit output per unit of wood in the orchard had become critical, as wood now cost a lot more to grow.

Changing Market Conditions

Preoccupation with the transition to denser orchards on clonal dwarfing rootstocks to produce greater and higher quality yields may have blinded American apple growers to a crisis emerging in the 1980s. The crisis, a severe drop in apple prices, was due to the over-production of two apple varieties that were losing popularity with consumers. The varieties, Red Delicious and Golden Delicious, accounted for the majority of apples grown in the United States by the mid-1980s. By this time one of every two apples produced was a Red Delicious and most of the remainder was Golden Delicious. The broad adaptability of Red and Golden Delicious to a range of growing conditions, their successful use on clonal rootstocks, the fire blight resistance of Red Delicious, and the abundant spur types and sports of both varieties all contributed to their popularity with growers. These factors, combined with the emergence of CA storage, ultimately led to a flooding of the market.

Advent of Controlled Atmosphere Storage

CA storage was developed in the 1960s and helped fuel the apple industry's over-production of Red and Golden Delicious. Using a gas-tight refrigerated room with most of the oxygen removed, CA storage slows down the rate of deterioration of fruit, allowing apples to be stored up to 10 months before flavor and quality diminish. CA storage enabled American growers to produce far more fruit than could be consumed in the harvest season and to store and release "fresh" apples onto the market in late winter and early spring. These apples would command a sufficiently high price to cover the cost of storage.

Decline of Red and Golden Delicious Apples

Throughout the 1970s and 1980s, consumer satisfaction with Red and Golden Delicious apples meant that most of the CA-stored crop would be purchased by late January. However by the mid-1980s, other competitors had lured consumers' attention away from these standard choices. First, new apple varieties originating in New Zealand, Japan, and Australia, namely Braeburn and Gala, Fuji, and Granny Smith, respectively, had entered the market. Secondly, new apple imports were reaching the United States in March and April from the southern hemisphere. Consumers immediately responded to the availability of new varieties, picking these over the time-worn varieties. In addition, nearly six-month old American CA-stored apples could not compete with one-month old apples imported from Chile, New Zealand, and South Africa. As a result, Red and Golden Delicious apples languished in CA storage and the value of the crop dropped dramatically (Phillips 04: n.p.).

American Apple Industry Crash

Red and Golden Delicious were ideal varieties from the growers' perspective, but they represented a corner of the market that the industry had backed into, causing a crash when the market expanded. As the price of Red and Golden Delicious apples plummeted to half the value of new competitors, growers responded by ripping out their Red and Golden Delicious orchards. The late 1980s and 1990s saw a wave of replacement of apple orchards throughout the United States, and the new plantings used the improved standard in rootstocks and tree spacing efficiency. The most clonal dwarfing rootstocks (on M. 9 or M. 26) were used for the new apple orchards, planted at the highest densities of 1,000 to 2,000 trees per acre. These rootstocks produced such weak aerial growth that the trees would need permanent support with stakes or trellises (Figure 4.10). Using the scion varieties, growers aimed to resurrect their businesses from the industry crash, in which more than 20 percent of American orchards went out of business (Phillips 04: n.p.).

Figure 4.10 *(above, left)* Photograph of a contemporary dwarf apple tree grafted onto an M. 9 clonal dwarfing rootstock, pruned into a spindle style (one main stem) and staked for support (by Bruce Barritt, courtesy of Washington State University).

New Apple Orchards

The adoption of more dwarf apple trees and more dense spacing on an array of support systems has continued. At the present time, 60 percent of all apples harvested in the United States are grown on the most clonal dwarfing rootstocks, using M. 9 (or a new virus-free Polish series P. 1, or the Russian Budagovsky B. 9) or M. 26. Currently, 90 percent of all apple trees in Washington state are grown on M. 9 (Barritt 07: interview). Of the remaining apple orchards throughout the country, 30 percent are grown on clonal semi-dwarfing rootstocks, i.e., M. 7, MM. 106. The last 10 percent use seedling rootstocks and generally grow apples for food processing rather than the fresh market (Barritt 01: interview) (Figures 4.11 and 4.12).

Figure 4.11: Photograph of a contemporary semi-dwarf apple orchard of free-standing, unstaked trees, grafted to M. 7 clonal dwarfing rootstocks, planted at 500 trees per acre, WA (S. Dolan, 2007).

Figure 4.12 Photograph of a contemporary dwarf apple orchard in eastern Washington, with trees grafted onto M. 9 clonal dwarfing rootstocks, trellised for support and planted at 1,000 trees per acre (S. Dolan, 2007).

Characteristics of Contemporary Pear Orchards

At the present time, the adoption of dwarfing rootstocks for other orchard species is not as complete as with the apple. The majority of pear orchards are still grown on Bartlett seedling rootstocks as they were before World War II, but they are now planted more densely and the size of the tree is controlled with summer pruning and the use of growth regulators. Young pear orchards are planted at 200 to 400 trees per acre, rather than 50 trees per acre in the former period. In California, where cold-hardiness is not a limiting factor, a large proportion of pear orchards are now grown on EM. Quince A or C clonal dwarfing rootstocks. In colder growing regions of the country and particularly in the Pacific Northwest, a pear orchard transition is underway from seedling rootstocks to clonal dwarfing rootstocks of the Old Home x Farmingdale (OH x F) series.

Advent of New Pear Rootstocks

The OH x F series is a patented group of clonal rootstocks developed in the 1980s by an Oregon nursery in partnership with Oregon State University (OSU). The rootstocks were derived from a hybrid between the Old Home and Farmingdale pear varieties, both discovered on an old homestead in Farmingdale, Illinois in the 1910s by OSU professor, F. C. Reimer. He discovered that both varieties were resistant to fire blight and "pear decline," a fatal disease that attacks the graft union between the scion and the rootstock. Reimer hybridized the varieties and began experimenting with their use as blight-resistant rootstocks. Today, the most promising members of the hybrid OH x F series, OH x F 333 and OH x F 51 are widely available to growers in nurseries throughout the United States. These rootstocks can be grafted to a range of pear varieties and will create a tree approximately two-thirds the size of a standard. They are considered to be semi-dwarf (Stebbins 95: 3) (Figure 4.13). Other new frontiers in pear orchard research include the selection of new varieties such as Red d'Anjou, Russet Bosc, and Red Clapp, and the development of training systems for high-density orchards such as trellises.

Characteristics of Contemporary Peach Orchards

Likc the pear, peaches are still predominantly grown on seedling rootstocks, but are also grown at higher densities than before World War II. A common peach rootstock is the Lovell Seedling, a seedling derived from the Lovell canning variety. While this is not a clonal rootstock, the Lovell Seedling does result in a slightly dwarfed peach tree when grown on wet or heavy soils. The use of high-density management systems for peach began in the 1970s when hedgerow, tilted tree, and palmate pruning systems were explored (Teskey 78: 275). These systems were found to yield twice as many fruits per acre as the traditional system using 15 x 20 feet spacing, without the need for dwarfing rootstocks. Today, new peach orchards are planted at a density of 500 trees per acre, at 10 x 8 feet spacing. The trees are maintained at less than 10 feet tall using up to two rounds of summer

Figure 4.13 Photograph of a research orchard of dwarf pears grafted to various OH x F rootstocks. The whitewashed trunks protect the trees from winter sun scald (courtesy of Oregon State University Mid Columbia Agricultural Research and Extension Center, 2004).

pruning and chemical growth regulators (Figure 4.14). Recent trends in peach varieties have replaced Elberta as the leading freestone peach variety for the fresh market. While Elberta is still grown, it has lost popularity due to mediocre quality and flavor and the tendency to drop before maturity. The variety has been superseded by earlier ripening and more red-colored varieties, such as Red Haven, Redskin, and Sun Crest (Teskey 78: 194).

Characteristics of Contemporary Cherry Orchards

The transition from seedling to clonal dwarfing rootstocks for cherry is underway in the cherry growing regions of the East, upper Midwest, and Pacific Northwest. The traditional seedling rootstocks for full-sized cherry trees are Mazzard for sweet cherries and Mahaleb for sour cherries, though either rootstock can be used for each species. Sweet cherries, with varieties such as Bing and Rainier, are derived from the species *Prunus avium*. Sour cherries, with Montmorency as the most important commercial variety, are derived from *Prunus cerasus*. Before 1970, Mazzard, the wild seedling of *P. avium*, was typically used as a seedling rootstock and would give rise to full-size sweet cherry trees, planted at 25 x 25 feet spacing. Mahaleb, an old variety of *P. cerasus*, when used as a seedling rootstock, creates a semi-standard sour cherry tree (slightly dwarf), and is planted at 15 x 15 feet spacing (Tukey 64: 164). Sour cherries are mechanically harvested as they are grown for food processing rather than the fresh market. Developed in the 1960s, mechanical harvesting equipment shakes each tree, causing the fruit to drop and be caught in a canopy positioned beneath each tree.

Figure 4.14: Photograph of a contemporary high-density peach orchard supported by trellis wires, TX (courtesy of Texas A & M University).

By contrast, higher quality sweet cherries are hand-picked, and so the development of clonal dwarfing rootstocks has focused more on rootstocks for sweet cherry trees than for sour. Most sour cherry orchards are still grown on Mahaleb rootstocks in the United States.

Advent of Cherry Dwarfing Rootstocks

During the 1960s, the East Malling and Merton Research Stations in England worked on the development of clonal dwarfing rootstocks for cherry. With less tendency to genetically mutate than the apple, progress was slow. Research made use of hybrids between the species *P. avium* and *P. cerasus* to generate genetic variability and yield new characteristics for rootstocks. Trials led to the selection of the rootstock "Colt" by East Malling and its release in the 1970s. The Colt rootstock gives rise to a semi-standard (slightly dwarf) sweet or sour cherry tree, approximately 25 percent smaller than standard. Colt was readily accepted in many areas of Europe and by the 1980s, had become the industry standard for both sweet and sour cherries. First planted in the United States in 1978, Colt's lack of compatibility with some varieties, sensitivity to drought and lack of cold hardiness, made it only viable in well irrigated, warmer areas of the country (ISHS Horticulture 468). The sweet cherry trees in the John Muir National Historic Site House Unit orchard were replaced in the 1980s and early 1990s with the sweet varieties Bing and Black Tartarian grafted onto Colt rootstocks. While these trees were typical California nursery stock by the late 1980s, the semi-standard trees did not accurately reflect the standard trees grown by John Muir before 1915 (Dolan 06: 44).

The need for other clonal dwarfing rootstock selections for cherry became an important area of research in the 1980s and 1990s in the United States and Europe. A more widely adaptable dwarfing rootstock was pursued for benefits beyond greater quality yield and lower costs of production. Other potential benefits included accelerating the earliness of bearing in sweet cherry (which bears fruit older than sour cherry), the potential to net the orchard for bird protection and the potential to confer disease resistance, particularly from prune dwarf and necrotic ringspot viruses. In the late 1980s, the "Gisela" series of clonal dwarfing rootstocks were introduced into the United States from Germany. As trials were being conducted by various research stations, the first plantings occurred in sweet cherry growing areas. Of the Gisela series, GI. 5 and GI. 6 currently appear to be among the most promising for commercial sweet cherry orchards. Hardier and more drought tolerant than Colt, GI. 5 creates a semi-dwarf tree half the size of the Mazzard seedling rootstock, while GI. 6 produces an even smaller dwarf cherry tree (www.goodfruit.com/archive, 7/1998) (Figure 4.15).

Figure 4.15: Photograph of a contemporary semi-dwarf Sweet cherry orchard with trees grafted onto clonal dwarfing rootstocks, showing closer spacing than full-size, standard trees on seedling rootstocks, WA (S. Dolan, 2007).

Trellised Orchards

New frontiers in cherry orchard research are concerned with optimization of the use of the new clonal dwarfing rootstocks. A new standard for high density cherry orchard spacing and training is sought that balances fruit size versus yield. For unlike apples and pears, clonal dwarfing cherry rootstocks dramatically increase yield but also reduce fruit size. A new "canopy architecture" is needed that balances the training system, pruning, fertilizing, and flower and bud thinning to maintain fruit size with high yields. The Tatura Trellis system, developed by Dutch horticulturist Bas Van Den Ende in New South Wales, Australia, is among the most sophisticated systems being examined in the United States. This system uses spindle trees (with only one or two branches) trained at acute angles on permanent trellises (Figures 4.16 and 4.17). Tree height is limited to eight feet and trees are planted two to five feet apart in rows six feet apart. With 2,000 trees per acre covered by a bird net, the Tatura cherry orchard has more in common with the appearance of a vineyard than the traditional orchard prior to World War II. Like the contemporary apple, pear, and peach orchard, the traditional spacing of the sweet cherry orchard has become archaic (Barritt 01: interview).

Contemporary Plum and Apricot Orchards

With plum and apricot, trends have moved toward the adoption of clonal rootstocks more than the use of clonal dwarfing rootstocks. The consistency of clonal rootstocks provide the benefits of greater uniformity in tree form, youthful age of fruit bearing (precocity), adaptability to growing conditions and disease resistance. Like peach, dwarfing rootstocks for plum and apricot were less important by the late 1900s, as these are naturally smaller tree species (Teskey 78: 365). In particular, Japanese plums (*Prunus salicina*) that are hand-picked for the fresh market are borne on even shorter trees than European plums (*Prunus domestica*), which are mostly dried for prunes or used in food processing. Processed plums are mechanically harvested with shake and catch equipment like sour cherries, and the equipment is more suited for use with slightly larger trees with taller trunks that provide room for the shaker attachment. However, with hand-picked Japanese and Japanese hybrid varieties, the trend is to adopt high-density systems, using hedgerows and bush forms to expedite tree management and harvesting.

Development of Plum and Apricot Rootstocks

The East Malling Research Station worked on the standardization of clonal rootstocks for plum between the 1930s and the 1960s, and made five selections from the Myrobalan seedling plum: "Myrobalan A, B, C, D ,and E" (Tukey 64: 202). Of these selections derived from the species *Prunus cerasifera* (Cherry plum), Myrobalan B has become the most widespread clonal rootstock for European plum in North America, and is the most common plum rootstock in the United States beside the

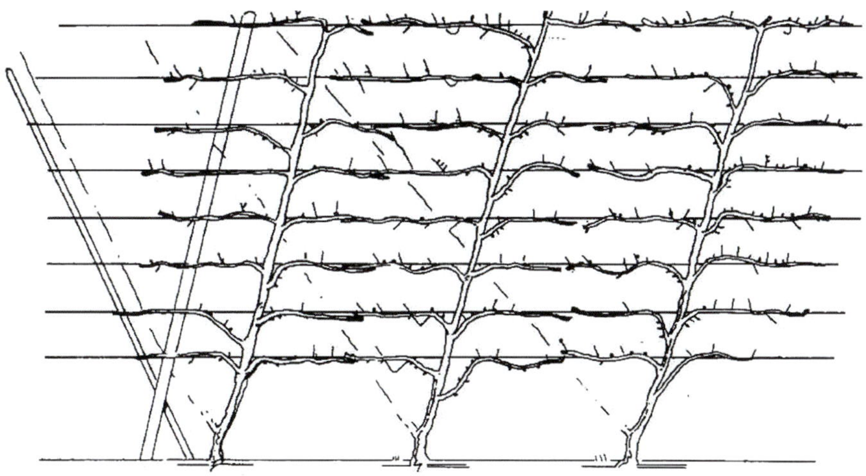

Figure 4.16: *(above)* Drawing of dwarf plum trees trained on a Tatura Trellis, using two main vertical spindles tilted to make a v-shape (rear spindle is shown by dashed line) (courtesy of the State of Victoria, Department of Primary Industries, Australia).

Figure 4.17: *(left)* Photograph of a contemporary dwarf Sweet cherry orchard, grafted onto clonal dwarfing rootstocks and trained on a Tatura Trellis. This is very likely to be the form of the orchard of the future, with 2,000 trees/acre, eastern WA (courtesy of Washington State University).

traditional Myrobalan seedling rootstock. Myrobalan B is compatible with both Japanese and European plum varieties and produces a vigorous, highly adaptable plum tree (Slingerland 02: n.p.). Plum orchards grafted to Myrobalan B rootstocks have been planted at 20 x 20 feet spacing. Myrobalan B, like the Myrobalan seedling rootstock, suckers freely from the roots, creating competition with the scion for water, nutrients, and light.

Since the 1970s, many other plum rootstocks have been explored and many are still under development. An improved selection from Myrobalan

C—Myrobalan 29C—is slightly less vigorous and less suckering than Myrobalan B, and is compatible with both Japanese and European varieties. Myrobalan 29C is now one of the most commonly used plum rootstocks in the Pacific Northwest and California along with Marianna 2624 (Reisner 06: n.p.). Marianna 2624 was selected in California in the 1970s as a hybrid between *Prunus cerasifera* and *P. munsoniana* (Munson plum). It is used as a slightly clonal dwarfing rootstock for both Japanese and European plums, and also for apricot (*Prunus armeniaca*). Marianna 2624 produces a semi-standard plum or apricot tree, approximately two-thirds the height of a standard tree. Unlike seedling Myrobalan or seedling apricot rootstocks, the Marianna 2624 rootstock is very tolerant of wet, heavy soils unlike seedling Myrobalan or seedling apricot rootstocks, and provides resistance to the prune black line virus. New plum and apricot orchards grafted to Marianna 2624 rootstocks are now planted as densely as 10 x 20 feet spacing.

In Europe, where horticultural land area is limited, more clonal dwarfing rootstocks are in use for Japanese and European plums and apricot. St. Julien A, a rootstock derived from *Prunus institia* (small-leaved European plum), creates a semi-dwarf tree, and related St. Julien K produces a dwarf plum tree. Both of these rootstocks were developed by East Malling and were recommended in the 1960s (Tukey 64: 209). In Europe, plum orchards grafted to St. Julien A are grown at 10 x 15 feet spacing and St. Julien K orchards are planted at 8 x 10 feet spacing. These and other high-yielding clonal dwarfing rootstocks are likely to further influence the evolution of American plum and apricot orchards towards greater density in the future.

"Plumcots" and "Pluots"

Besides clonal rootstocks and higher density systems, new frontiers in plum and apricot development include the hybridization of these closely related species. Belonging to the same subgenus *Prunophora*, plum and apricot genes can be combined to form a "plumcot," consisting of 50 percent plum and 50 percent apricot, or a "pluot," made of 75 percent apricot and 25 percent plum. No significant change in the complement of plum varieties has occurred since World War II (Table 4.1).

Apricot Varieties

While apricot varieties have also not changed a great deal since World War II, the once most popular variety, Moorpark, has lost ground to another old variety, Blenheim. Blenheim now accounts for 80 percent of all apricots grown in the United States. Other, less important cultivars include the old variety Tilton, and also the newer Wenatchee, Earlicot, and Autumn Royal (Reisner 06: n.p.).

Contemporary Citrus Industry

Since World War II, the United States has become one of the top 10 world producers of all important citrus fruits, namely: sweet oranges (*Citrus sinensis*), tangerine/mandarin (*C. reticulata*), grapefruit (*C. paradisi*), lemon (*C. limon*), and lime (*C. aurantifolia*). The United States is the world's leading producer of grapefruit and the second largest producer of orange, next to Brazil. Florida has more orange, grapefruit, tangerine, and lime orchards than any other state, due to the preference of these species for tropical growing conditions. California has the greatest number of lemon orchards, due to this crop's better adaptation to a Mediterranean climate (Reisner 06: n.p.).

Most Important Plum Varieties in 1945 to Present Period			
European Plum (Prunus domestica)	**Distinguishing Characteristics**	**Varieties**	**Use**
Green Gage Group	Green or golden, round	Green Gage Reine Claude Imperial Gage Hand	Dessert fruit or canning
Yellow Egg Group	Large, yellow, oval	Yellow Egg Golden Drop	Canning
Lombard Group	Large, red or pink, oval	Lombard Pond Victoria	Dessert fruit or canning
Prune Group	Dark blue or purple, oval	French Italian Stanley President	Drying
Japanese Plum (Prunus salicina)	**Distinguishing Characteristics**	**Varieties**	**Use**
	Medium to large, orange to red, round	Santa Rosa Burbank Shiro Beauty	Dessert fruit or canning

Table 4.1: The most abundantly grown European Plum (Prunus domestica) and Japanese Plum (Prunus salicina) varieties since World War II, with characteristics and uses (S. Dolan, 2007).

Advent of Citrus Dwarfing Rootstocks

During the second half of the 20th century, horticultural research in many countries, including the United States, Australia, Argentina, China, Spain, and Brazil, has produced a number of clonal dwarfing rootstocks for citrus. Of these, clonal rootstocks derived from the trifoliate orange (*Poncirus trifoliata*) are among the most commonly used for all citrus, with the exception of lime, which is already a shorter tree species (Figure 4.18). The trifoliate rootstock, Rubidoux, is used to dwarf sweet orange, tangerine/mandarin and grapefruit, to create a semi-dwarf tree, which is approximately half the size of the standard on seedling sweet orange rootstocks. The trifoliate rootstock, Flying Dragon, is used to dwarf the lemon to less than one-third the size of the standard tree on seedling rootstock (Tukey 64: 218).

Characteristics of Contemporary Citrus Orchards

Similar to other orchard fruit species, with the exception of the apple, the adoption of clonal dwarfing rootstocks for citrus is not as widespread as the adoption of high-density management systems. The benefit of clonal rootstocks is perceived in mostly the conferring of disease resistance and earlier fruit bearing. In the last several decades, the taller citrus tree species of grapefruit, sweet orange, and tangerine have been planted at tighter spacing and mechanically pruned into hedges. Rather than the traditional 30 x 30 feet spacing (40 trees per acre) planted before World War II, grapefruit is now grown at 20 x 25 feet spacing (100 trees per acre), and sweet orange and tangerine are grown at 15 x 20 feet spacing (140 trees

Figure 4.18: Photograph of the orange orchard at John Muir National Historic Site showing historically incompatible dwarf orange trees on the right, mixed with accurate standard orange trees on the left, CA (S. Dolan, 2004).

per acre). Lemon and lime are grown at 12-15 x 18-20 feet spacing with up to 180 trees planted per acre. The mechanical pruning system maintains the mature citrus trees as a continuous hedge with a width of 12 feet. The pruning machinery flat-tops the hedge at 12 feet tall (Reisner 06: n.p.). The appearance of the round-canopied, broadly spaced citrus orchard prior to World War II has been superseded.

Global Market Influence

The trend towards high density and the use of clonal or clonal dwarfing rootstocks has been the pattern for most orchard species in the years following World War II to the present. However, as with orchard history since 1600, the horticultural development of the apple in the last 60 years has outpaced other fruit species. Still, evolution of apple orchards in the United States lags behind Europe, Asia, and Australia, where new developments serve as a barometer for the future of all commercial fruit industries in the United States and globally. In the new emerging paradigm for commercial orchards, the concept of "global" is a central, controlling, but change-propagating force. This new global orchard paradigm proposes that orchards with full-sized, standard fruit trees bearing 19th-century varieties with wide spacing are landscapes of the past.

Advent of Patented Varieties

A harbinger of the new global orchard paradigm came in 1997, when New Zealand first successfully patented the apple variety Pacific Rose®. Originating in New Zealand, the patented variety could be licensed in other countries, where growers would pay license fees and royalties on their profits. Two years later, Jazz® and Pink Lady® were released, followed by Honeycrisp™ and Zestar!® by the University of Minnesota (Phillips 04: n.p.). The proprietary nature of these varieties reflected the global forces of competition in the marketplace during the 1990s.

After recovering from the late 1980s crash of Red and Golden Delicious prices, American growers faced a 40 percent increase in world apple production between the mid-1990s and the present. During this time China became the world's largest apple producer, growing one-third of the world's crop and exporting large amounts of low-priced apple juice concentrate. Overall, global apple production has grown more than 250 percent since the 1960s. Today the world produces enough apples to provide each person with 20 pounds per year, grossly outweighing demand (Phillips 04: n.p.). By the late 1990s, longtime American commercial orchards were once again uncompetitive, albeit this time in the global market. Their choice was to pay to adopt the new varieties, or convert to growing for the processed market.

Influence of Private Varieties

New Zealand's innovation was an attempt to counter the overproduction of unrestricted, cheap, common apple varieties through the development of private varieties. These new varieties, created through public and private partnerships, are valuable intellectual property that can be protected with patents, plant breeder's rights and trademarks. The value of the patented property can be controlled by limiting production, stipulating growing conditions, requiring a marketing strategy and guaranteeing quality. The owner of a variety can refuse to license to a grower with an incompatible climate, too short a growing season, or a poor business plan. Growers will have to give up some control of their planting practices and horticultural methods to buy into a consortium of licensees. However, the new global marketplace views this interdependence and discipline as a recipe for high prices and success (Barritt 01: interview).

High-density orchards with tightly spaced trees on clonal dwarfing rootstocks are a key partner to the patented variety concept. The precocity of dwarf trees and their short life span are expected to help growers keep pace with variety obsolescence in the future. As predicted by Harold Tukey Sr. in 1964, consumers have been found to grow tired of common varieties and express a preference for new choices. In the global marketplace, researchers have calculated that variety obsolescence occurs every 10 years, prompting the need to accelerate the breeding of new varieties.

21st Century Orchard Model

In the new global orchard paradigm, as new patented varieties are released every decade, high-density orchards will be rotated like row crops. Dwarf trees will be torn out and replaced with more dwarf trees of a new variety. The new dwarf trees will bear fruit in the second year, providing a profit for nine of the ten-year lifespan of the orchard. The high-density orchard is currently considered to be the optimal model for a short rotational system to combat variety obsolescence, and patenting new varieties is regarded as a feasible way to control value (Barritt 01: interview) (Figures 4.19 and 4.20). Orchard horticultural research in the United States is repositioning to charter this new course for apple and other orchard species, all but ensuring archaic uniqueness of the old orchards in the national park system.

Uniqueness of Archaic NPS Orchards

The more orchards change, the more distinguished the orchards dating prior to World War II become, and the more unique the experience of an orchard landscape of the 1940s or earlier will be for park visitors. After tracing the history of orchards in the United States from 1600 to the present, it is evident that American orchards have been transformed through the centuries, and will continue to change in the future. Orchards have always been a reflection of societal values and economic and technological realities, and they have been made to fit the changing realities. The many historic orchards in national parks and elsewhere are cultural landscapes that memorialize these events, trends, and eras in American history. As we preserve orchards that are 50 years of age or older and that retain significance and physical integrity, their cultural resource value will continue to grow in importance. Genetic biodiversity conservation combines with visitor education as potential societal benefits. However, these very old living organisms are in decline and the need for preservation intervention is growing more urgent.

Figure 4.19: Photograph of a contemporary high-density apple orchard grafted onto clonal dwarfing rootstocks and trained on a Tatura Trellis, eastern WA (J. Dolan, 2007).

Figure 4.20: Photograph of a contemporary high-density apple orchard trained on a Tatura Trellis, eastern WA (J. Dolan, 2007).

Summary

The trend towards higher density, dwarf fruit tree orchards in the period from World War II to the present was fueled by the need to lower costs of production in an increasingly competitive marketplace. The development overseas of clonal rootstocks (genetically cloned, standardized rootstocks) influenced American horticultural practices by exposing American growers to more sophisticated horticultural technologies. The discovery by European researchers that select dwarfing rootstocks could produce greater yields of higher quality fruit than seedling rootstocks led to the development of clonal dwarfing rootstocks. First developed for apple before World War II, clonal dwarfing rootstocks were then created for pear, plum, cherry, apricot, and citrus. These rootstocks provided multiple benefits for growers, including more quality fruit per unit of wood production, earlier fruit bearing, disease resistance, and easier orchard management.

By the 1960s, semi-standard M. 7 and M. 106 clonal rootstocks had been adopted for apple orchards throughout the United States, with tree spacing transformed from 30 x 30 feet (40 trees per acre) before World War II to 10 x 18 feet (240 trees per acre). By the late 1980s, new apple orchards were planted on the most clonal dwarfing rootstocks M. 9 and M. 111, requiring staking or training systems to support the tree. Accompanying the use of clonal dwarfing rootstocks was the adoption of high-density management systems, using trellises to grow dwarf spindle trees at 2-5 x 6-10 feet spacing (1,000 to 2,000 trees per acre). Following suit, clonal dwarf rootstocks for pear, plum, apricot, and citrus have been discovered since World War II and adopted since the 1970s. Only peach does not have an accepted clonal dwarfing rootstock. With all orchards, the mass planting of singular varieties (with the exception of rows of pollenizer trees) became the norm for management efficiency, resulting in vast monoculture orchards over thousands of acres.

By the late 1970s, pear, peach, plum, apricot, and citrus were grown at tighter spacing, with hedge systems being explored in concert with clonal dwarfing rootstocks for fresh market fruits. Hedge systems had very close tree spacing within the rows, and wider spacing between the rows, though still tighter spacing than before World War II. Hedge systems co-evolved with mechanical pruning systems invented during the 1960s. Mechanical harvesting systems, also developed in the 1960s, promoted less density in orchards grown for food processing to provide space for the "shake and catch" equipment.

Gradually, the movement towards higher density management systems affected all orchard fruits that would be hand-picked for the fresh market. Fruit varieties grafted to clonal rootstocks rather than clonal dwarfing rootstocks would also be planted at tight spacing, and be maintained as compact trees by summer pruning and newly developed chemical growth regulators (Table 4 .2).

Other trends included the adoption of spur-type mutations of several apple and pear varieties and the popularity of red sports of apple and pear varieties between the 1950s and the 1970s. Development in the 1960s of Controlled Atmosphere storage fueled the overproduction of Red and Golden Delicious apple varieties. Their ultimate loss in popularity and value in the 1980s was a response to new global competition. New varieties and new imports devalued Red and Golden Delicious and stimulated a trend towards the growing of a broader range of new apple varieties. A new round of peach varieties was developed in the 1970s, with earlier fruiting and more red-color characteristics, superseding the dominant Elberta variety. The range of pear, plum, apricot, and citrus varieties did not expand a great deal in the late 20th century, with early 20th century commercial varieties continuing to dominate.

Table 4.2: Typical spacing of fruit orchards in the 1880-1945 period and at present, indicating a dramatic increase in the density of orchards since World War II (S. Dolan, 2007).

Change in Orchard Spacing Between 1945 and the Present

Fruit Species	Historic Orchard Spacing 1880 – 1945 (feet)	Historic Number of Trees per Acre	Contemporary Orchard Spacing (feet)	Contemporary Number of Trees per Acre
Apple	30 x 30	40	10 x 18 (M. 7 / MM. 106) 2-5 x 6-10 (M. 9 / MM.111)	240 1000–2000
Apricot	20 x 20	100	10 x 20	200
Cherry	25 x 25	60	2–5 x 6–10	1000–2000
Grapefruit	30 x 30	40	20 x 25	80
Lemon / Lime	20 x 20	100	12–15 x 18–20	140–180
Orange / Tangerine	30 x 30	40	15 x 20	140
Pear	30 x 30	40	12 x 20	180
Peach	15 x 20	140	8 x 10	500
Plum	20 x 20	100	10 x 20	200

At the turn of the 21st century, global market forces are shaping the work of orchardists and horticultural researchers. The new frontier appears to be the patenting of varieties that will be licensed to growers and be superseded every 10 years by other new patented varieties. New orchards will be grown on clonal dwarfing rootstocks in high-density management systems, which will produce the greatest quantity of highest quality fruit for the majority of the 10-year life span of the orchard. These new rotational orchards are planned to keep pace with the concept of variety obsolescence.

As a result of the trends in orchard history since World War II, earlier orchards with widely spaced trees on seedling rootstocks have come to represent archaic horticulture. Thousands of varieties have been lost as a result of the decreased number of varieties grown, and the new patented varieties are distinct from those of earlier periods. These changes have distinguished the older orchards in national parks dating prior to World War II, which represent earlier periods in the history of American horticulture. Historic orchards in national parks and elsewhere are now the repositories of rare varieties or strains of varieties, and are becoming rare examples of extant old fruit tree forms and layouts. Orchards have changed radically since World War II, and the rate of change can be expected to continue to grow.

Part II:

Technical Information for Registering Orchards in the National Register of Historic Places

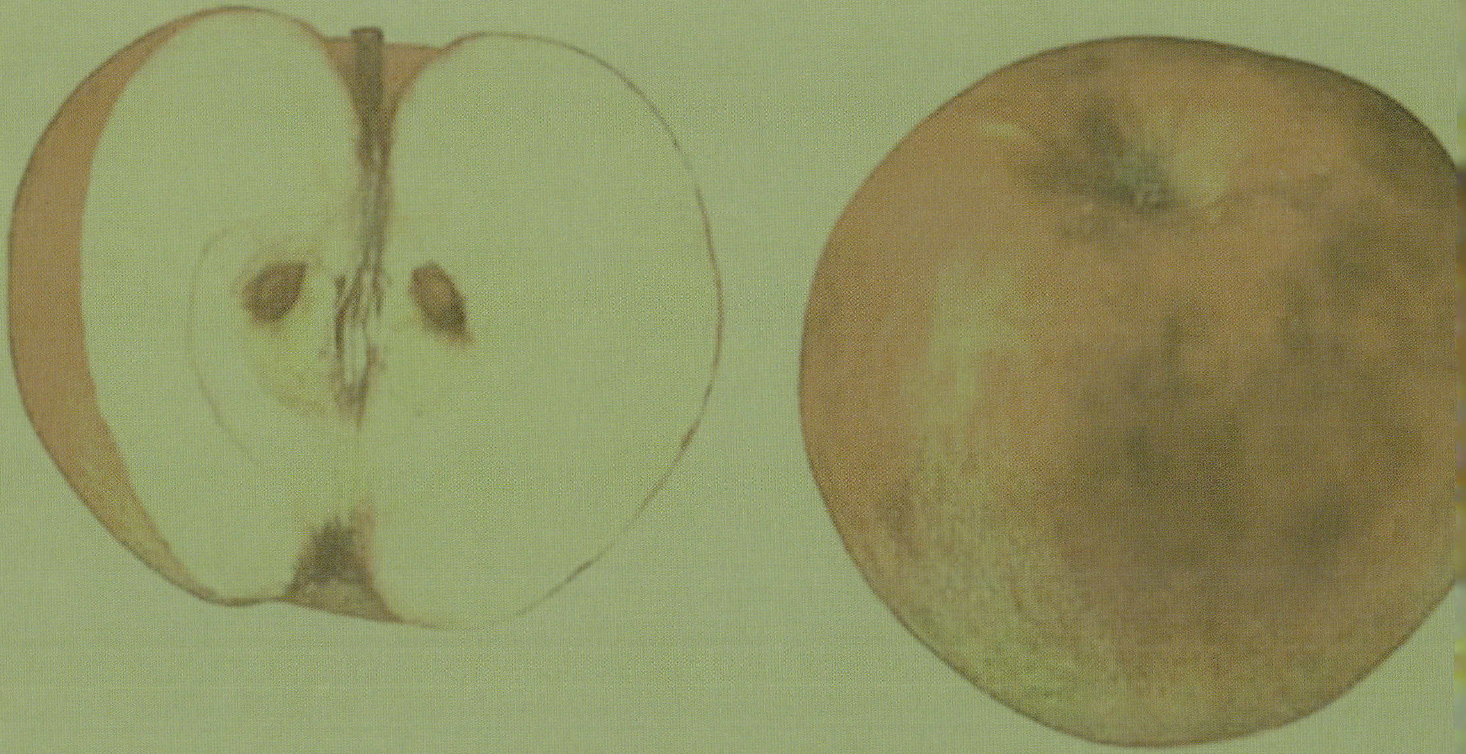

Chapter 5
Evaluating the Significance and Integrity of Historic Orchards and Fruit Trees

Orchards and Fruit Trees as National Register Property Types

This chapter provides guidance on using the National Register criteria to evaluate the significance and integrity of historic orchards and fruit trees. The chapter is intended to serve as a supplement to *National Register Bulletin 15: How to Apply the National Register Criteria for Evaluation.* While these registration requirements and examples focus on National Park Service (NPS) cultural resources, the requirements apply equally to non-NPS properties.

Historic orchards, a group of fruit trees, or a single fruit tree, may be found eligible for listing in the National Register of Historic Places. Orchards or fruit trees may be eligible for listing individually, or as a contributing feature that is part of a larger historic property. Like other cultural landscapes, orchards may be listed individually on the National Register as historic districts or historic sites. A group of fruit trees or a single fruit tree may be listed individually as a historic site. If orchards or fruit trees lack individual distinction but contribute to the significance and integrity of a larger property, then an orchard, group of fruit trees, or a single fruit tree may be included in a National Register nomination as a contributing feature to a historic district or a historic site (Table 5.1).

This chapter provides definitions of the National Register property types—historic district, historic site, and contributing feature—and their eligibility requirements. Although it is beyond the scope of this document, additional property types and registration requirements may be associated with the historic context of orchards, such as historic nurseries, horticultural societies, horticultural publishers, and horticultural libraries. A multiple property submission format may be suitable for these associated property types.

Table 5.1: Table showing the application of National Register Property Types to orchards and fruit trees (S. Dolan, 2007).

Listing Orchards and Fruit Trees on the National Register of Historic Places		
Resource	**National Register Property Type**	
	Listed Individually	***Listed as Contributing***
Orchard	Historic District or Historic Site	Feature
Group of Fruit Trees	Historic Site	Feature
Single Fruit Tree	Historic Site	Feature

Definitions of "orchard" and "fruit tree"

As indicated in preceding chapters, orchards have existed in a variety of forms since the 1600s. But regardless of when it originated, any orchards is a horticultural system centered upon a plantation of woody trees of fruits or nuts. Banana, pineapple, palms, and other non-woody or monocotyledonous commercial fruits are excluded from this definition. The plantation may have been raised from seed or from young trees planted out, and may have a regular geometry or no geometry at all depending on its period, history, and growing conditions. An orchard may be a complex horticultural system with a number of landscape characteristics and features (described in "Defining Integrity" later in this chapter) or be a relatively simple system with few landscape characteristics.

A complex orchard is a horticultural system consisting of a plantation of trees of one or a number of species, and one or a number of varieties. A complex orchard may have a complex spatial organization, such as having multiple blocks of fruit trees of various spacing, and may have a number of associated use areas, such as pasture for grazing livestock to provide manure for the orchard, fruit and equipment storage areas, and residential and garden areas. A complex orchard may also have a circulation system; an irrigation system, an array of buildings and structures, a cover crop for the orchard floor, and a boundary system for browse or wind protection, or property delineation.

A simple orchard is a horticultural system consisting largely of a small plantation of trees of one or numerous species, and one or numerous varieties, and a ground cover. Historically, farm orchards were five acres in size or less, which was an adequate size to supply a farm family for a year. Commercial orchards were typically larger than five acres, sized to raise large quantities of fruit for sale and consumption elsewhere. Commercial orchards have been typically more complex horticultural systems than farm orchards, largely due to their scale of operation. However, farm orchards may have more or less complexity.

Fruit trees are distinguished from orchards in this chapter in order to emphasize that orchards are horticultural systems and fruit trees are not. Fruit trees may exist in small groups where they were deliberately planted, or may be irregularly distributed, as they are the remnants of a fragmented, former orchard. In either case, the extant trees cannot be identified as an orchard, but rather as a planting or a remnant orchard. In both cases, the fruit trees are not intact horticultural systems (or never were, in the case of small plantings) and do not have a complement of landscape characteristics.

Defining the Significance of Orchards and Fruit Trees

Like other cultural resources, historic orchards and fruit trees are a combination of tangible and intangible features, qualities, and values. To be eligible for listing in the National Register of Historic Places, an orchard, group of fruit trees, or a single fruit tree must possess significance in at least one of four aspects of cultural heritage. More commonly, orchards and fruit trees are part of larger historic properties and are not listed in the National Register for their own significance, but as contributing features to a historic district or historic site. To have their own significance, however, an orchard or group of fruit trees must not only be associated with something noteworthy in the past, but the association must be manifested in physical substance as defined in the *National Park Service Cultural Resource Management Guideline* (NPS CRM 97, 9). In other words, an individually eligible orchard or group of fruit trees must have both significance and integrity. An orchard's physical substance, like other cultural landscapes, can be described by its landscape characteristics: the processes and patterns on the land that are the tangible evidence of the activities and habits of the people who occupied, developed, and shaped the land to serve human needs.

Defining the significance of an orchard or fruit trees requires a thorough understanding of the history and existing conditions of the resource in relation to its associated historic context. A broad historic context is provided by the first four chapters of this document, tracing the important events, trends, and patterns in the national history of orchards from 1600 to the present. A more specific regional or local historic context may be needed to analyze and evaluate the significance of a specific orchard or a group of fruit trees. With knowledge of the national, state, or local historic context, the history of the specific orchard or fruit trees, and the existing conditions, the significance can be defined and the landscape characteristics and features that contribute to significance can be identified. The significance of the overall property must be identified at one of three levels: national, state, or local.

A property is significant on the local level when its historic context represents an aspect of the history of a town, city, county, cultural area, or region. The level of significance is defined by the importance of the property within the historic context, not by the physical location of the property. For example, if a type of orchard is found throughout a state, or extends over two states, but its importance relates only to a particular county, the orchard property would be considered of local significance.

A property is significant on the state level when its historic context represents an aspect of the history of the state as a whole. These properties do not necessarily have to belong to a type of orchard found throughout the entire state, but can be located in only a portion of the state's present political boundary. It is the property's historic context that must be important statewide.

A property is significant on the national level when its historic context represents an aspect of the history of the United States as a whole. Properties designated as nationally significant and listed in the National Register are the prehistoric and historic units of the national park system and those properties that have been designated National Historic Landmarks. A property with national significance provides an understanding of the history of the nation by illustrating the nationwide impact of events, persons associated with the property, its type or style, or information potential. The orchard must be exceptional in representing the theme of the history of orchards in the nation. Nationally significant properties do not necessarily have to belong to a property type found throughout the entire country, but can be located in only a portion of the present political boundaries. It is their historic context that must be important nationwide.

National Register Criteria for Evaluation	
Criterion	**Type of Significance**
A.	Associated with events that have made a significant contribution to the broad patterns of our history
B.	Associated with the lives of persons significant in our past
C.	Embodying the distinctive characteristics of a type, period, method of construction, or that represent the work of a master, or that possess high artistic values, or that represent a significant and distinguishable entity whose components may lack individual distinction
D.	Having yielded or may be likely to yield, information important in prehistory or history

Table 5.2: The National Register Criteria for Evaluation indicate types of significance embodied by properties eligible for listing on the National Register (source: National Register Bulletin 15: How to Apply the National Register Criteria for Evaluation).

As defined by the National Historic Preservation Act of 1966 and the National Register criteria, a cultural landscape is eligible for the National Register if it possesses the quality of significance in American history, architecture (including landscape architecture and planning), archeology, engineering, and culture. Orchards and fruit trees may possess the quality of significance in any of these areas: American history, landscape architecture, archeology, and culture (horticulture is covered by each of these areas). Like other cultural landscapes, an orchard or group of fruit trees must be shown to be significant for one or more of the following National Register Criteria for Evaluation (Table 5.2).

Applying the National Register Criteria to Orchards and Fruit Trees

The National Register Criteria for Evaluation, identified in Table 5.2, can be applied to orchards, groups of fruit trees, or single fruit trees, and may be applied in more than one way. Three of the four criteria, specifically A, B, and C, have two or more applications to orchards and fruit trees. The different applications of each criterion are outlined in a table as numbered categories. The following narrative then identifies each criterion category and illustrates the category with examples of historic orchards and fruit trees in the national park system.

Table 5.3: Table showing the application of National Register criterion A to orchards or fruit trees as three categories, A1 to A3 (S. Dolan, 2007).

Applying National Register Criterion A to Orchards and Fruit Trees	
Criterion	**Type of Significance**
A.	Associated with events that have made a significant contribution to the broad patterns of our history
Category A^1	The orchard or fruit trees have played an important role in prehistory, in the settlement history, or in the subsequent history of development of an area
Category A^2	The orchard or fruit trees are associated with a historic horticultural innovation, practice or event
Category A^3	The orchard or fruit trees are associated with a historic event not related to horticulture

Category A^1

The orchard or fruit trees have played an important role in prehistory, in the settlement history, or in the subsequent history of development of an area.

The earlier chapters in this document illustrate how orchards or fruit trees often played a major role in the settlement history of an area, first as subsistence crop plants for food and beverages, and later as cash crops and aesthetic objects of desire. Native fruit trees such as some plum species and nuts were cultivated as subsistence crops by indigenous peoples in prehistoric times. Post contact with Europeans and Euro-Americans, indigenous peoples obtained non-native orchard fruit seeds and plants, and the planting and cultivation of fruit trees were among the European influences that altered their subsistence patterns. The suitability of a particular area for orchard fruit production influenced local and regional settlement patterns, and allowed for the settlement of a canyon floor in arid desert land, the settlement of steeply mountainous terrain in temperate regions, or the acquisition of a homestead patent in western lands. The significant property under this criterion may consist of an orchard or a system of orchards, or may contain an orchard that is part of a larger rural property. As described in the "Defining Integrity" section, where an orchard or group of fruit trees contributes to the significance of a larger property, the orchard or tree group may have less integrity to convey the significance of the historic context than an orchard or fruit trees that are individually significant under this criterion.

Category A[1] Examples

In Capitol Reef National Park, the Fruita Rural Historic District contains 65 acres of historic orchards with approximately 2,500 trees. The orchards are directly associated with the first settlement of this Utah canyon by Mormons in the late 1800s and early 1900s. The unique microclimate of the canyon, in combination with Fremont River water available for irrigation, enabled a wide variety of orchard fruits to be cultivated in this arid region, allowing for the development of a discrete Mormon community (Figure 5.1).

In Canyon de Chelly National Monument, where the Navajo have farmed the canyons for 300 years, seeds of non-native peach were introduced by the Spanish to the Navajo in the late 1700s. This allowed for the cultivation of peaches in the canyon washes, altering the subsistence practices of some Navajo farmers. Peach orchards and groupings of peach trees found in the canyons of the monument today are associated with this proto-historic change in Navajo subsistence practices.

In Yosemite National Park, disparate groups of seedling fruit trees are the remains of homestead orchards, associated with the federal government's disposition of western lands through the Homestead Act of 1862. The seedling trees were sown as a minimal investment method for converting uncultivated land to agricultural use, fulfilling one legal requirement for a homestead patent. The extant seedling trees are directly associated with Euro-American settlement of the area through the influence of the Homestead Act.

Category A[2]

The orchard or fruit trees are associated with a historic horticultural innovation, practice or event.

As explained in the four earlier chapters of this document, orchards and fruit trees have changed a great deal from 1600 to the present due to horticultural innovation, changing practice, or influential events. Orchards or fruit trees in the national park system and elsewhere may be directly associated with horticultural innovation, or manifest the influence of this innovation, practice or event. As described in the "Defining Integrity" section, where an orchard or group of fruit trees contributes to the significance of a larger property, the orchard or tree group may have less

Figure 5.1: Site plan of the Fruita Historic District in Capitol Reef National Park showing the influence of the Fremont Valley and sheltering canyon walls on the settlement and planting of the valley in orchards by Mormons in the late 1800s. Today, the district retains 65 acres of orchards, UT (from Fruita Historic District Cultural Landscape Report, 1997).

integrity in order to convey the significance of the historic context than an orchard or fruit trees that are individually significant under this criterion.

Category A² Examples

In Lake Chelan National Recreation Area of the North Cascades National Park Service Complex, the Buckner Homestead Historic District contains a 90-year old apple orchard that reflects a revolutionary event in the modern period of orchard fruit growing: the discovery of the Delicious apple variety by the Stark brothers. This discovery transformed the growing of apples in the 20th century. Among the oldest trees of the Buckner orchard are trees of the Common Delicious strain of the Delicious variety—Stark Brothers Nursury's name for the early, unimproved variety they acquired as "Hawkeye," a red and yellow apple. The Buckner orchard is the oldest and largest known plantation of the variety in the United States and represents the advent of the Delicious variety before it was turned into a red apple, or Red Delicious variety, through strain selection in the 1920s (Figure 5.2).

In the Moses H. Cone Memorial Park of the Blue Ridge Parkway, the orchards are associated with the beginning of the modern era of commercial orchards, which began in the late 1800s with developments that led to improved methods of production. The orchards, planted between 1899 and 1901, contain low-headed trees, a late 19th century innovation to promote fruit tree management. They also contain recently invented pesticide spraying equipment for pest control and an apple barn, with insulation and a ventilation system for improved fruit storage. The characteristics of these

Figure 5.2: Photograph of the Buckner Orchard in North Cascades National Park Service Complex, WA, showing a low-headed apple orchard with Common Delicious (Hawkeye) variety trees, dating from the 1920s (S. Dolan, 1999).

orchards are unique to the beginning of the modern period of orchard fruit growing, and are a direct result of late 19th century innovation. The Cone orchards are some of the only early modern-era examples remaining in the United States (Figure 5.3).

Category A^3

Orchards or fruit trees associated with a historical event not directly related to horticulture.

Between 1600 and the Great Depression of the 1930s, orchards and fruit trees were commonplace in agricultural landscapes, and until 1900 most Americans had farming lifestyles. Whether planted as a commercial operation or purely for home use, the majority of farms had fruit trees during this period. During the Civil War, historians have noted that orchards were among the most common "signatures" of farms, along with wheat and cornfields. With orchards and fruit trees omnipresent in America's rural landscapes and lifestyles, they were inevitably the settings for historic events not directly related to horticulture, such as Civil War battles. Today many orchards and fruit trees in the national park system and elsewhere are associated with a historical event not directly related to horticulture, or the cultivation of the fruit trees themselves. In terms of conveying the significance of historic context, an orchard or group of fruit trees that is significant for the location of an event may have less integrity than a property that's significant for events associated with horticulture, such as the settlement of an area (category A^1) or a horticultural innovation (category A^2).

Figure 5.3 Photograph showing a view of the Flat Top Orchard from the mansion house at the Moses Cone Estate of the Blue Ridge Parkway, dating from 1899, NC (S. Dolan 2001).

Category A^3 Examples

In Sleeping Bear Dunes National Lakeshore, a group of fruit trees of mixed cider varieties are associated with illicit cider-making activities during the event of Prohibition in the 1920s. Planted in a discrete, narrow draw, hidden from view, the 80-year-old group of trees is directly associated with the inflated, black-market value of cider resulting from the federal law prohibiting the sale, manufacturing, and transportation of alcohol. The law was ratified as the 18th Amendment of the U.S. Constitution in 1919, and was rescinded as the 21st Amendment in 1933. This site is one of the country's most intact examples of a Prohibition Era group of fruit trees planted for illegal cider production (Figure 5.4).

At Manzanar National Historic Site, groups of apple and pear trees are associated with the event of Japanese-American internment during World War II. Planted and abandoned by the Owens Valley Improvement Company in the 1920s, the fruit trees were rehabilitated by Japanese-American internees of the Manzanar War Relocation Center, and brought into productive use (Figure 5.5). Horticulture occupied the time of many internees during their confinement at Manzanar. Fruits and other crop plants were cultivated for subsistence, and small, ornamental gardens were created near the center of the camp. The groups of apple and pear trees at Manzanar are contributing features within the larger historic property of the former internment camp, and their significance is directly associated with the historic context of the property.

Figure 5.4: *(below, left)* Photograph of a hidden cider apple orchard at Sleeping Bear Dunes National Lakeshore planted during the 1920s Prohibition Era to produce illegal cider. A field visit was part of the NPS Historic Orchard Preservation Workshop held at the park in 2001, MI (C. Goetcheus, 2001).

Figure 5.5: *(below, right)* Photograph of a low-headed pear orchard and staff at Manzanar National Historic Site. The trees were planted by Owens Valley settlers in the early 1900s. The land became an internment camp for Japanese Americans during World War II, when these and other orchards were tended and harvested by internees, CA (S. Dolan 2006).

Table 5.4: Table showing the application of National Register criterion B to orchards or fruit trees as two categories, B1 to B2 (S. Dolan, 2007).

Applying National Register Criterion B to Orchards and Fruit Trees

Criterion	Type of Significance
B.	Associated with the lives of persons significant in our past
Category B^1	Orchards and fruit trees are associated with a person or persons who played an important role in horticultural history, or in the horticultural development of the area
Category B^2	Orchards and fruit trees are associated with a historically significant person not directly related to horticulture, such as a political figure, writer or artist

Category B^1

Orchards and fruit trees are associated with a person or persons who played an important role in horticultural history, or in the horticultural development of the area.

In the evolution of American orchards and fruit trees from 1600 to the present, on the national, state, or local level, particular individuals have contributed to the development of American horticulture through scientific discovery, academic scholarship, economic and technological innovation, and craftsmanship. Some orchards or fruit trees in the national park system and elsewhere are associated with these individuals, as the subject of their experimentations or as the product of their labors. As described in the "Defining Integrity" section, where an orchard or group of fruit trees contributes to the significance of a larger property, the orchard or tree group may have less integrity in order to convey the significance of the historic context than an orchard or fruit trees that are individually significant under this criterion.

Category B^1 Examples

At John Muir National Historic Site, a late 19th century pear orchard is associated with Dr. John Strentzel, John Muir's father-in-law who planted the trees (Figure 5.6). The orchard was one of hundreds of acres of orchards planted by Dr. Strentzel, a medical doctor who first settled and developed the land with commercial orchards. Dr. Strentzel is an important figure in California's horticultural history on the local level, being one of the first commercial orchardists in the Martinez area. Strentzel was active in the

Grange, a social and political organization of farmers and growers, and a local leader in horticultural practice through teachings and writings. He was also the builder of the area's first fruit depot for the transport of growers' fruit to San Francisco.

In Sleeping Bear Dunes National Lakeshore, Frank Farm on North Manitou Island is associated with the Stark brothers, developers of the Red and Golden Delicious apple varieties that transformed commercial orchard practice in the 1900s (Figure 5.7). Frank Farm was owned by William Stark, one of the Stark brothers. Frank Farm's extensive commercial orchards helped support the Stark brothers' operations.

Figure 5.6: Photograph of the pear orchard at John Muir's gravesite in John Muir National Historic Site, planted in the early 1880s by Muir's father-in-law Dr. John Strentzel, a regionally renowned horticulturist, CA (S. Dolan, 2004).

Figure 5.7: *(above, left)* Photograph of the vast Frank Farm orchard on North Manitou Island, Sleeping Bear Dunes National Lakeshore owned by orchardist William Stark, one of the brothers of Stark Brothers Nursery, who developed the Red and Golden Delicious apple varieties, MI (C. Goetcheus, 2001).

Figure 5.8: *(above, right)* Photograph of the orchard at Adams National Historical Park associated with Presidents John Adams and John Quincy Adams, both gentleman farmers and orchardists, MA (S. Dolan, 2001).

At Adams National Historical Park, the orchard is associated with four generations of the Adams family, from President John Adams in 1787, his son President John Quincy Adams, his grandson Charles Francis Adams, to his great grandson Brook Adams, in 1927 (Figure 5.8). Both presidents were gentlemen farmers interested in experimentation and the development of horticultural methods, and their successors appreciated the family orchards and continued their cultivation and maintenance. While the oldest trees of the orchard date only to Charles Francis Adams' tenure, the orchard has remained on the same site since the time of John Adams, and contains varieties and tree forms that represent all periods of the four Adams' generations.

Category B^2

Orchards and fruit trees are associated with a historically significant person not directly related to horticulture, such as a political figure, writer or artist.

Orchards or groups of fruit trees are more commonly associated with significant persons not related to horticulture than with persons significant in horticulture. As described in the "Defining Integrity" section, properties not significant for horticulture but for their association with a historically significant person not related to horticulture, may have less integrity as an orchard or group of fruit trees to convey the significance of the historic context than properties significant for horticulture, such as under criterion B, category B^1.

Category B^2 Examples

The orchards at John Muir National Historic Site are associated with the naturalist John Muir. Muir inherited his father-in-law's commercial orchards in the 1880s and used their revenues to support his family and his conservation-related activities. His world-renowned activities included founding the Sierra Club, advocating for the creation of national parks, and supporting the creation of a National Park Service. The oldest fruit trees at the site were managed by John Muir and the landscape as a whole is dominated by orchards, as it was in Muir's time.

At Eugene O'Neill National Historic Site the walnut orchard is associated with the 20th century playwright Eugene O'Neill (Figure 5.9). O'Neill wrote several plays while living on his ranch property in Danville, California. He found both inspiration and solace in the bucolic setting of the ranch. The ranch contained a number of orchards during O'Neill's time, as it does today, though the walnut orchard is the oldest and most accurately represents an orchard of the period. The walnut orchard is a feature of the larger ranch landscape, which is significant for its association with O'Neill.

Figure 5.9: Aerial photograph taken in 1951 of orchards at the home and ranch of American playwright Eugene O'Neill, now the Eugene O' Neill National Historic Site, CA (courtesy of the park archives).

Table 5.5: Table showing the application of National Register criterion C to orchards or fruit trees as two categories, C1 and C2 (S. Dolan, 2007).

Applying National Register Criterion C to Orchards and Fruit Trees

Criterion	Type of Significance
C.	Embodying the distinctive characteristics of a type, period, method of construction, or that represent the work of a master, or that possess high artistic values, or that represent a significant and distinguishable entity whose components may lack individual distinction
Category C^1	Orchards or fruit trees that embody the distinctive characteristics of a type, period, horticultural system or style, or contain a rare or unusual genotype, such as a variety or strain of a variety, or feature continuity of traditional use and occupancy
Category C^2	Orchards or fruit trees that were part of a historic designed landscape; the orchard was designed for research, or for the demonstration of "good" horticulture

Category C^1

Orchards or fruit trees that embody the distinctive characteristics of a type, period, horticultural system, or style; or contain a rare or unusual genotype, such as a variety or strain of a variety; or feature continuity of traditional use and occupancy.

Between 1600 and the present, the distinctive characteristics of orchards and fruit trees in the United States changed considerably, with evolving horticultural practices and cultural values. The appearance of orchards and fruit trees can be distinguished in four different periods, described in the first four chapters of this document. Orchards and fruit trees in the national park system date from all four periods, though only the first three are historically significant periods, as they gave rise to horticultural systems, tree types, or styles that are now archaic and increasingly rare. Orchards or fruit trees dating from the first three periods may be eligible for listing in the National Register under Criterion C if they retain integrity and are associated with a significant historic context.

Distinguishing Characteristics

Characteristics that distinguish orchards and fruit trees in each period include: tree genotype; tree form; pruning style; tree layout; tree spacing; orchard size; and orchard type. The evolution of these characteristics is described in the previous four chapters, and is summarized in Table 5.6. Other associated characteristics and features include irrigation systems, fruit storage facilities, pest control, and cover crops. These are discussed under "Defining Integrity" and "Landscape Characteristics." Orchards and fruit trees dating from the first three horticultural periods (1600-1800, 1801-1880, and 1881-1945) can be significant for their distinctive orchard and landscape characteristics under Criterion C.

The four periods in the history of orchards and fruit trees are also distinguishable by the types of fruit varieties grown, with some varieties being extremely common in one period and almost absent from the next. With the general trend toward increasing rarity or extinction of varieties from the earlier periods, orchards and fruit trees in the national park system and elsewhere may contain rare or unusual genotypes that are among the last surviving members of that genotype. A genotype is a unique genetic signature, and each fruit variety or fruit variety strain (a subset of a variety) has a unique genetic signature.

Fruit varieties are man-made creations that are the product of several thousand years of hybridization and selection in the case of apple, pear, and olive, and several hundred years of hybridization and selection in the case of cherry, citrus, and nuts. These genotypes are rendered extant through human intervention in vegetative propagation. Cultivated varieties cannot be reproduced from seed, and will become extinct if not perpetuated through asexual or vegetative reproduction. Orchards and fruit trees of rare or unusual varieties, or early examples of varieties subsequently modified by many strains, are significant for their distinctive characteristics of genotype, where the variety or strain had significance within a historic context at the national, state, or local level.

As described in the "Defining Integrity" section, properties significant for horticulture under Criterion C, category C^1, must retain sufficient integrity to convey the significance of the historic context. More integrity of horticultural characteristics may be required and than properties not associated with horticulture. To be eligible for individual listing in the National Register under Criterion C, category C^1, an orchard or group of fruit trees must retain the horticultural characteristics that convey the significance of the historic context. This may include tree genotype, tree form, pruning style, tree layout, tree spacing, orchard size, and orchard type.

Table 5.6: Table identifying orchard design characteristics in different historic periods (S. Dolan, 2007).

Orchard Characteristics Through Time

Characteristic	Variables	Sub-variables	Examples	Period
Tree Genotype	Seedling		Ungrafted tree grown from seed	1600 – 1800 1801 – 1880
	Species		e.g., Apple, pear, cherry, etc.	All periods
	Variety		e.g., McIntosh apple; Bosc pear, Black Tartarian cherry, etc.	1801 – 1880 1881 – 1945 1946 – Present
		Variety strain	e.g., Starkrimson Red Delicious, etc.	1946 – Present
Tree Form	Ungrafted		Seedling tree	1600 – 1800
	Grafted		Variety tree	1801 – 1880 1881 – 1945 1946 – Present
		Standard	Full size tree with seedling rootstock	1801 – 1880 1881 – 1945
		Spur-type	Semi-standard strain of variety with many fruiting spurs	1946 – Present
		Dwarfed	Semi-standard; semi-dwarf; dwarf; naturally occurring dwarf or grafted to clonal dwarfing rootstock	1946 – Present
		Location of graft union	Below ground	1801 – 1880
			At ground level	1881 – 1945
			Above ground level	1881 – 1945 1946 – Present
			High on trunk	1946 – Present
		Type of rootstock	Seedling	1801 – 1880 1881 – 1945
			Clonal and clonal dwarfing (semi-standard, semi-dwarf, dwarf)	1946 – Present
Pruning Style	Unpruned		No removal of crossing branches or scaffold development	1600 – 1800 1801 – 1880
	Pruned		Scaffold development	1801 – 1880 1881 – 1945 1946 – Present
		Scaffold Style	Tall trunk, 4 – 8 ft before branching	1801 – 1880
			Short trunk, low-headed 1.5 – 3 ft before branching	1881 – 1945
			Tall trunk (proportionately) on short tree, 2 – 3 ft before branching	1946 – Present
		Pruning Style	Central leader; pyramidal; modified central leader	1881 – 1945 1946 – Present

Characteristic	Variables	Sub-variables	Examples	Period
Tree Layout	No geometry		Irregular layout	1600 – 1800
	Regular geometry		Grid or polygonal layout	1801 – 1880 1881 – 1945 1946 – Present
		Shape of regular geometry	Grid	1801 – 1880 1881 – 1945
			Quincunx	1881 – 1945
			Rectangle	1881 – 1945 1946 – Present
Tree Spacing	No regular spacing		Seedling orchard	1600 – 1800
	Regular spacing		Seedling or variety orchard	1801 – 1880 1881 – 1945 1946 – Present
		Regular spacing between rows	e.g., 30 feet between rows	Varies from 1801 – Present (see Table 4.2)
		Regular spacing within rows	e.g., 20 feet within rows	Varies from 1801 – Present (see Table 4.2)
Orchard Size	Farm orchard		< 5 acres	1600 – 1800 1801 – 1880 1881 – 1945
	Commercial orchard		> 5 acres – thousands of acres	1801 – 1880 1881 – 1945 1946 – Present
Orchard Type	Farm orchard		Homestead orchard	1600 – 1800 1801 – 1880 1881 – 1945
			Hospital Orchard/Poor Farm	1881 – 1945
	Fruit garden		Kitchen garden/Walled garden/ Door yard	1600 – 1800 1801 – 1880 1881 – 1945
	Commercial orchard	Nursery	Multiple species, multiple varieties; or single species, multiple varieties	1801 – 1880 1881 – 1945 1946 – Present
		Family orchard enterprise	Multiple species orchard and multiple varieties; less common – single species and a single variety	1801 – 1880 1881 – 1945
		Corporate orchard enterprise	Single species orchard and single variety/multiple varieties, less common – multiple species and multiple varieties	1881 – 1945 1946 – Present

Category C[1] Examples

At Delaware Water Gap National Recreation Area, the seedling apple trees at Roberts Farm are more than 200 years old, and represent the earliest period of fruit tree culture in America—1600-1800. This is the period before variety trees were used in farm orchards and nearly every American farm had seedling apple trees to provide cider, the subsistence beverage for the family. Seedling farm orchards and fruit trees were characterized by the large size of trees, tall tree trunks, unpruned tree form, abundant and poor quality of fruit, great longevity, and lack of geometry to tree layout (Figure 5.10). The group of apple trees at Roberts Farm is a rare example of the earliest American fruit tree culture, and is among the oldest surviving group of apple trees in the country.

In Yosemite National Park, the Curry Village Orchard in Yosemite Valley represents the second period of fruit tree culture in America, from 1801 to 1880. These apple trees, planted in the late 1850s, represent the 19th century style of fruit tree cultivation, characterized by a wide range of variety trees in one orchard, tall tree trunks, unpruned tree form, quality fruit, great longevity of trees, and geometric tree layout. The Curry Village Orchard is a rare example of the second period of American fruit tree culture and is one of the oldest extant apple orchards in the country, despite its use as a parking area since the 1930s (Figure 5.11).

Figure 5.10: *(above)* Photograph of an over 200-year-old seedling apple tree at Roberts Farm, Delaware Water Gap National Recreation Area displaying the characteristic 18th-century form of a tall trunk, very large tree size, and "wild" forest-like form, NJ (C. Pepper, 2007).

Figure 5.11: *(right)* Photograph of mid-1800s variety apple trees in the Curry Village Orchard, Yosemite National Park displaying the characteristic form of a tall trunk, large tree size, and relatively unpruned scaffold, CA (S. Dolan, 2006).

The Buckner Homestead Historic District orchard of Lake Chelan National Recreation Area (Figure 5.12), the Fruita Rural Historic District orchards at Capitol Reef National Park, the Frank Farm orchard at Sleeping Bear Dunes National Lakeshore, all represent the third or modern period in the history of fruit tree culture in America, from 1880 to 1945. These orchards, all planted between 1900 and 1945, share the similar characteristics of a narrow range of variety trees, short trunks or low-headed trees, pruned tree form (pyramidal or open bowl), quality fruit, medium longevity, and wide geometry of tree layout. They are rare examples of the third period of American fruit culture, and are among the most extensive, intact orchards with these distinctive characteristics in the country.

At San Juan Island National Historical Park in Washington State, English Camp contains a group of pear trees that is the remains of a late 19th- and early 20th century orchard. Two of the trees are rare varieties, White Doyenne and Belle Angevine syn. "Pound" pear. The varieties were accurately identified through DNA fingerprinting by the USDA National Plant Germplasm Repository for pears in Corvallis, Oregon. The 80- to 100-year-old trees are associated with the late 19th century homesteading by a British emigrant, after the Royal Marines and U.S. Army left the island following the resolution of an international boundary dispute (Figure 5.13).

Figure 5.12: Photograph of low-headed apple trees in the Buckner Orchard at North Cascades National Park Service Complex displaying characteristically short trunks, large canopy, open-bowl pruning style and wide spacing, WA (S. Dolan, 1999).

White Doyenne is an ancient and world-renowned variety that came from France via Italy. There it was described in 1550 as the old Roman pear, "Sementinum." White Doyenne was introduced to France in 1559, then to England and America, and now there are only two known sources of the variety left in the country (Seed Savers Exchange 2001: 281). The pear was a treasured favorite of American settlers with European heritage and was regarded as a collector's variety.

Figure 5.13: Photograph of the pear orchard at English Camp, San Juan Island National Historical Park, which has two rare varieties and dates to the late 1800s and early 1900s. Park staff and Western Washington Fruit Growers Association volunteers were collecting scionwood for germplasm conservation, WA (C. Teague, 2005).

The Belle Angevine (Pound) variety is another one of the oldest known pear varieties, dating to 17th century England. This heavy pear can weigh over three pounds. With no commercial qualities, the variety was only useful for subsistence. It does not soften and cannot be eaten raw, but can be stored all winter (even on the tree) and is excellent for baking. The American Pomological Society dropped the variety from its fruit catalog in 1909, due to its lack of commercial viability. The variety was relatively common in more remote farm orchards of Euro-American settlers until the late 19th century, for its over-wintering subsistence value. Today, there is only one known nursery source for the variety in the United States (Seed Savers Exchange 2001: 267).

The group of fruit trees at English Camp is significant under Criterion C, category C^1, for its rare varieties that were once important in the United States among pear orchardists and homesteaders. These rare varieties are biologically significant, but are also historically significant because they convey the historic context of the horticultural and homesteading efforts of a British emigrant homesteader on San Juan Island in the late 19th century.

At Moses H. Cone Memorial Park on the Blue Ridge Parkway, in 1989, one of the late 19th-century orchards was discovered to contain a tree of the variety Gragg. One of 75 apple varieties planted by Moses Cone's grounds superintendent, F. L. Mulford, in 1899 and 1900, Gragg is now a rare variety in the United States. Mulford considered Gragg a worthy commercial variety in North Carolina at the end of the 19th century when the state was renowned for locally originating and distributing hundreds of apple varieties. The Moses Cone tree was discovered by Dr. Elwood Fisher of James Madison University during the preparation of the Moses H. Cone Memorial Park Cultural Landscape Report. Before his death, Dr. Fisher maintained one of the largest collections of fruit trees in the nation, with more than 1,100 varieties of apple. Dr. Fisher obtained a cutting of the Gragg tree to add to his collection. Other than at the Moses Cone Estate and in the Fisher Collection of Harrisonburg, Virginia, the variety has only one other known source in the United States (Seed Savers Exchange 2001: 89). The significance of Gragg in the Moses Cone orchards as a rare variety under Criterion C, category C^1, is one of several types of significance of the historic property. The Moses Cone orchards are significant for the Gragg variety because it conveys the historic context of North Carolina enterprise in apple variety development in the late 19th century.

Category C²

Orchards or fruit trees that were part of a historic designed landscape; the orchard was designed for research, or for the demonstration of "good" horticulture.

From the late 1700s to the early 1900s, orchards were incorporated into the design of large country estates or designed landscapes, as both aesthetic and productive components. Eighteenth-century texts such as Batty Langley's *Pomona, or the Fruit Garden Illustrated,* and 19th-century texts such as Andrew Jackson Downing Jr.'s *A Treatise in the Theory and Practice of Landscape Gardening,* promoted the inclusion of orchards or fruit gardens in designed landscapes and even provided graphic examples of landscape plans with orchards (Figures 5.14 and 5.15). The orchard was regarded as a highly aesthetic, ornamental landscape feature or space that decorated the landscape with beautiful blossoms, fine fruits, and shady walks or rides within allées of green boughs. A well tended orchard of choice fruit

Figure 5.14: Historic drawing proposal for a mid-1800s country place estate showing orchards incorporated into the overall landscape design (from Copeland 1866, courtesy of Dinsmoor and Company).

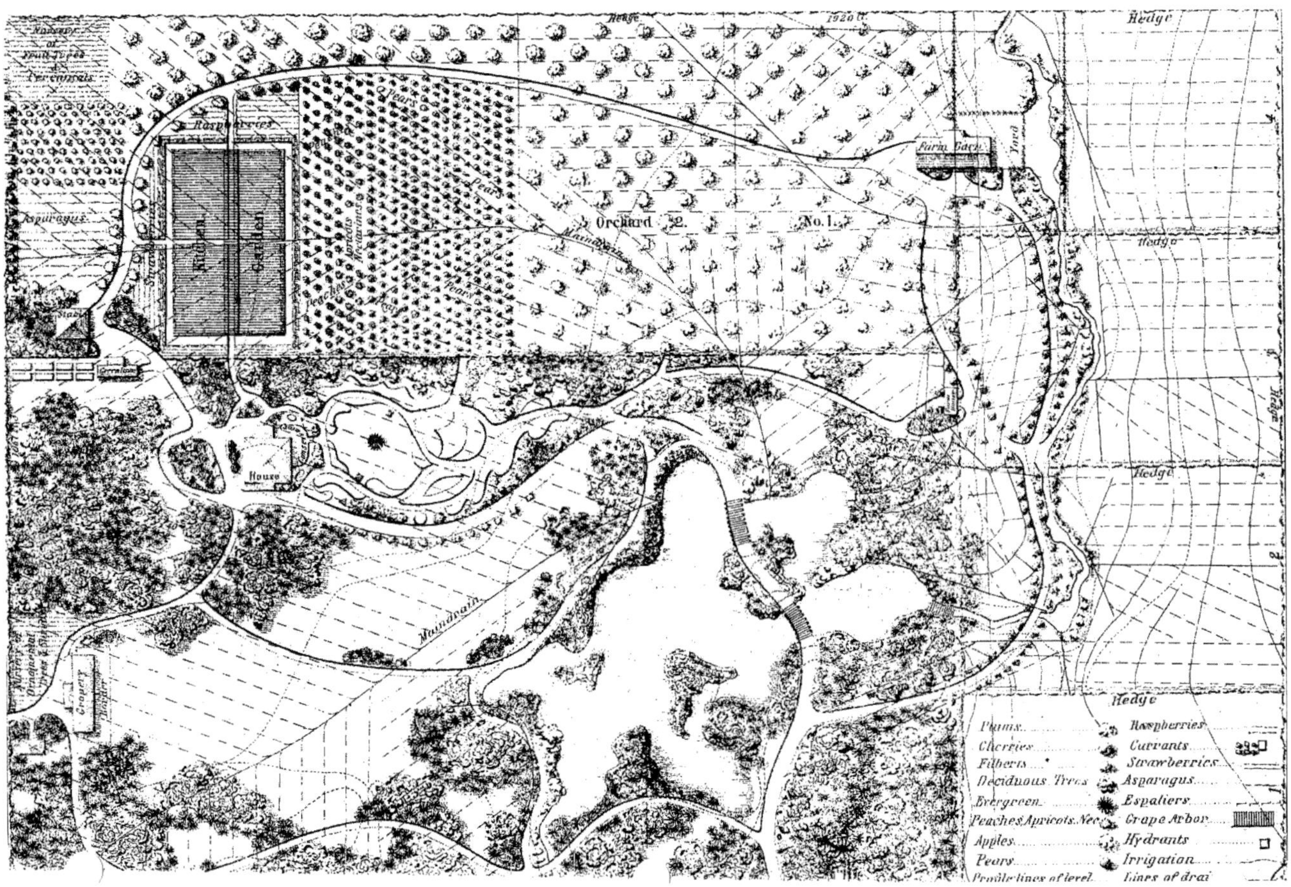

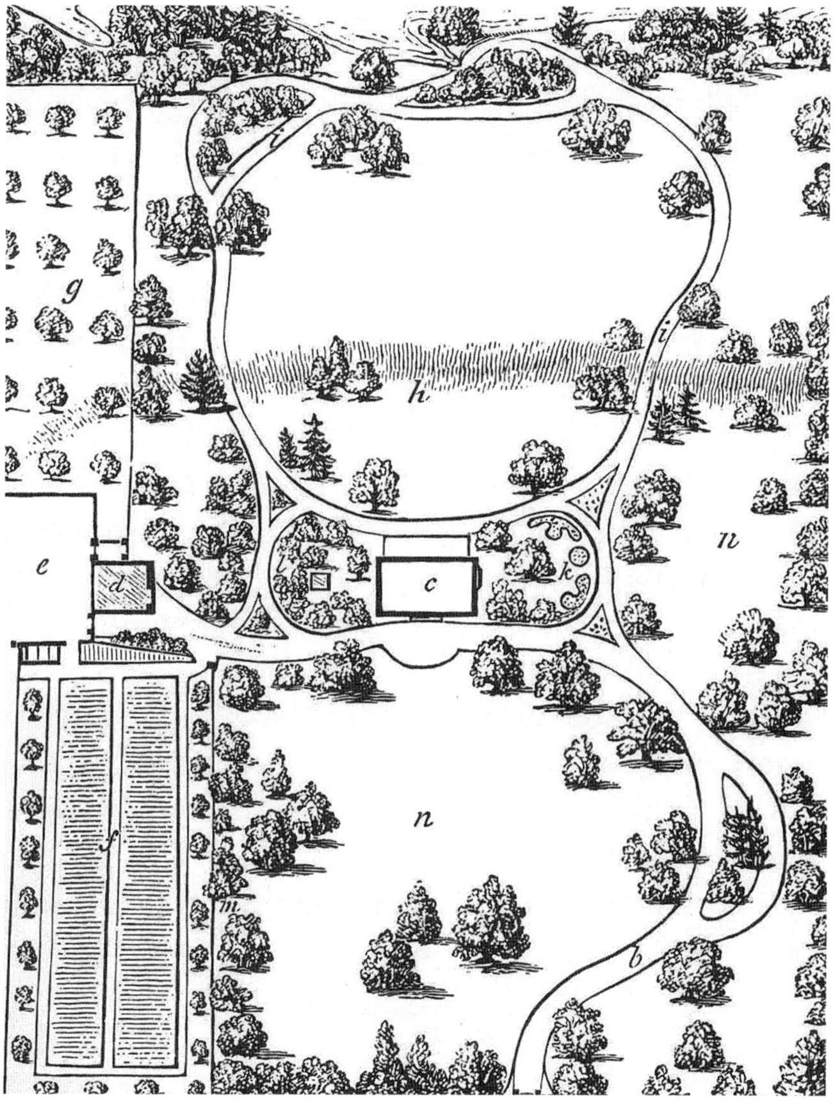

Figure 5.15: Historic drawing proposal for a mid-1800s gentleman's farm, Tudor Farm, designed by Andrew Jackson Downing, with a fruit garden (f) and orchards (g) incorporated into the overall landscape design (from Downing 1852, courtesy of J. Wiley Co.).

varieties was a hallmark of the owner's sophistication. Where an orchard or group of fruit trees contributes to the significance of a larger property, such as a historic designed landscape, the orchard or tree group may have less integrity in order to convey the significance of the historic context than an orchard or fruit trees that are individually significant under this criterion.

Category C^2 Examples

At Moses H. Cone Memorial Park of the Blue Ridge Parkway, orchards are part of a historic designed landscape that was laid out by Moses Cone in the late 1800s and early 1900s. A wealthy industrialist, Cone designed a system of carriage roads around views of orchards, mountains, lakes, forests, and pasture. Cone had more than 260 acres of orchards densely planted

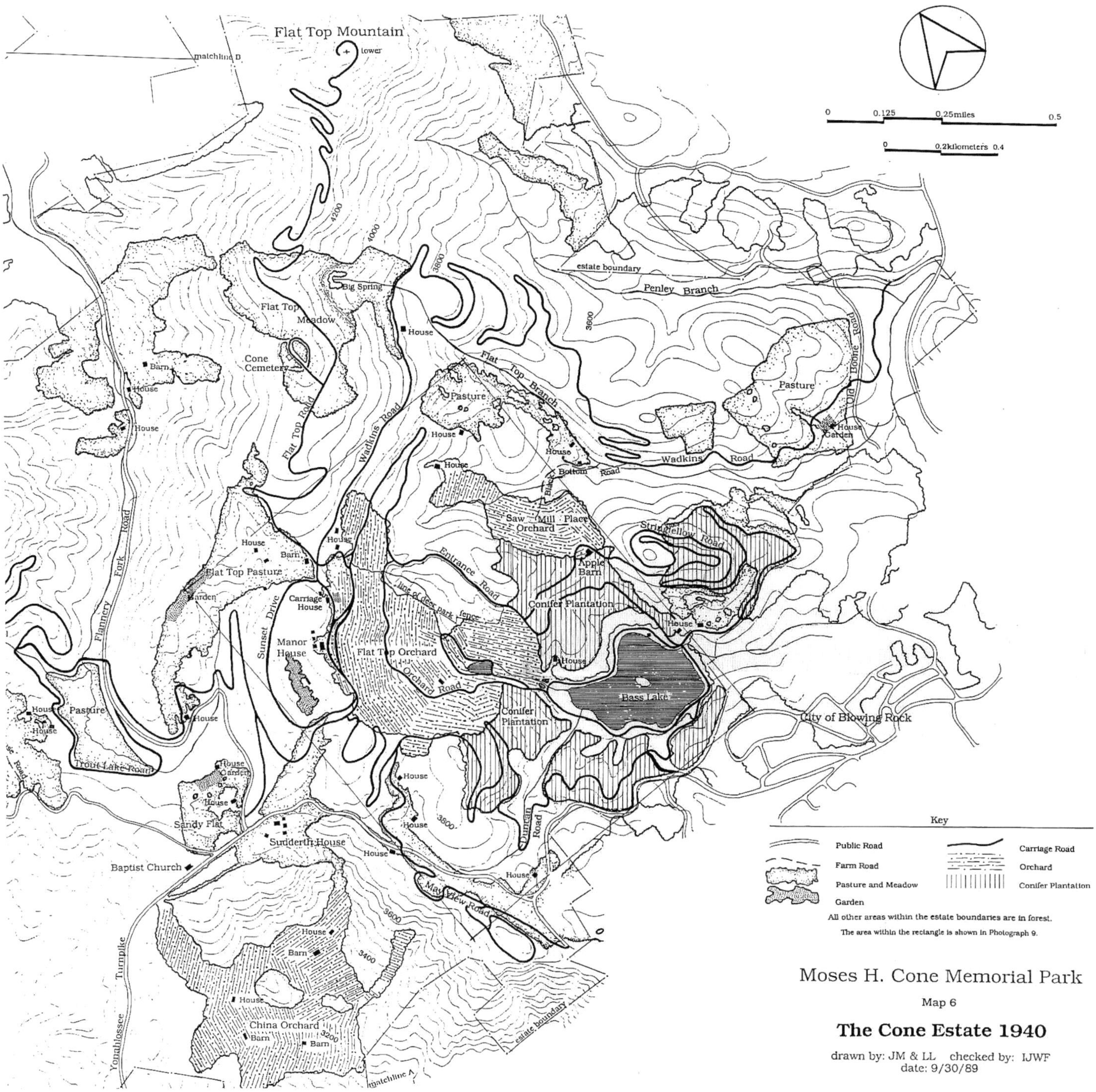
Flat Top Mountain
tower
matchline B
0 0.125 0.25miles 0.5
0 0.2kilometers 0.4
estate boundary
Penley Branch
Big Spring
Flat Top Meadow
House
Cone Cemetery
Barn
Flat Top Road
Wadkins Road
Flat Top Branch
Pasture
Old Boone Road
Garden
Wadkins Road
Bottom Road
Saw Mill Place Orchard
Stringfellow Road
Apple Barn
Entrance Road
Conifer Plantation
line of deer park fence
Flat Top Pasture
Carriage House
Sunset Drive
Manor House
Flat Top Orchard
Orchard Road
Bass Lake
City of Blowing Rock
Flannery Fork Road
Trout Lake Road
Sandy Flat
Sudderth House
Baptist Church
Duncan Road
Mayview Road
China Orchard
Yonahlossee Turnpike
estate boundary
matchline A
3600
3400
3200
Key
Public Road
Farm Road
Pasture and Meadow
Garden
Carriage Road
Orchard
Conifer Plantation
All other areas within the estate boundaries are in forest.
The area within the rectangle is shown in Photograph 9.
Moses H. Cone Memorial Park
Map 6
The Cone Estate 1940
drawn by: JM & LL checked by: IJWF
date: 9/30/89

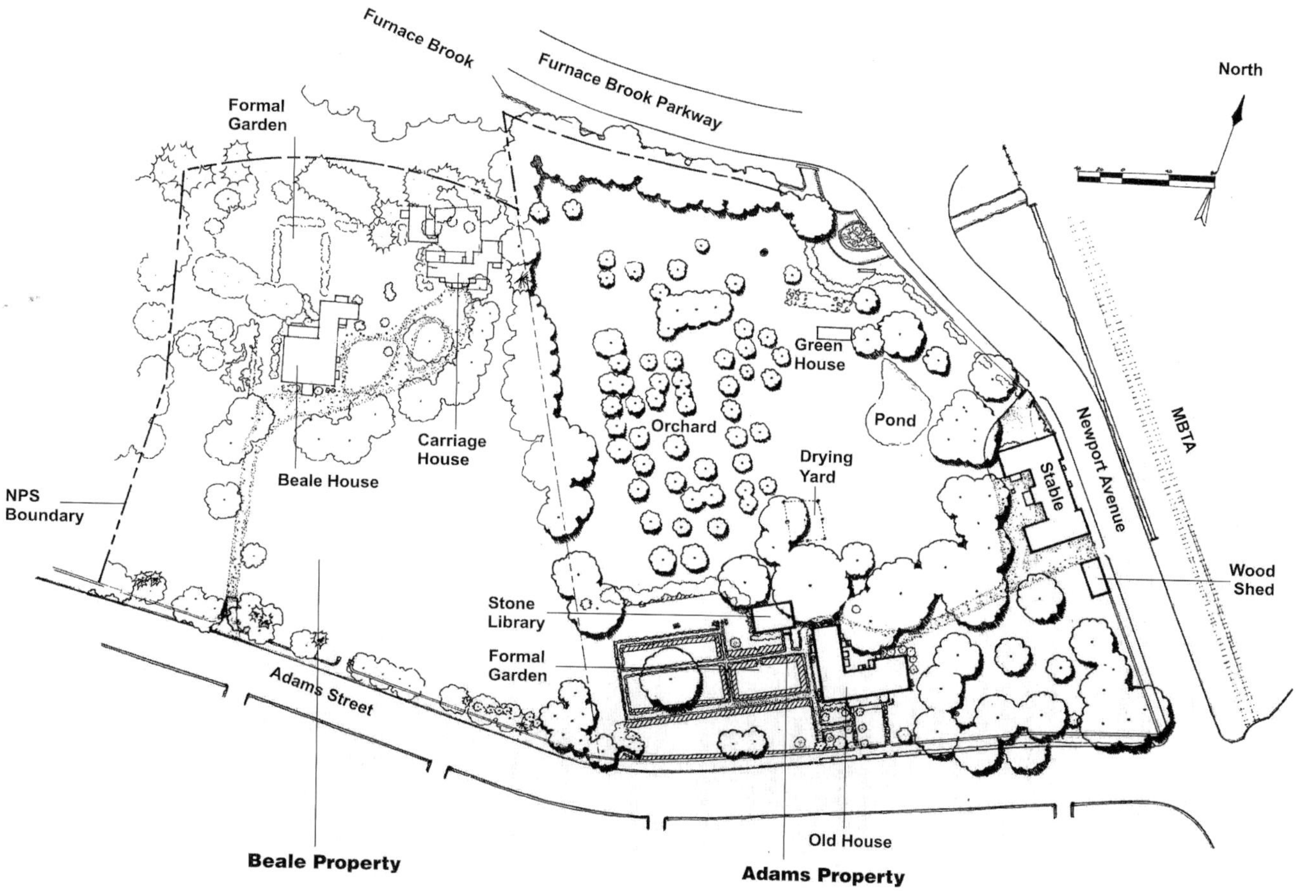

with 29,000 trees (Figure 5.16). He designed his carriage roads to present composed scenes of orchards, lakes, and forested mountains. The Moses Cone orchards are contributing features within a larger historic property.

At Adams National Historical Park, the orchard is part of a designed historic landscape that has its origins in an early 18th century formal estate of a wealthy Boston merchant, Major Leonard Vassall. His designed landscape contained a formal garden and orchard adjacent to the estate house, similar to the design of English country estates of the period. John Adams acquired the property in 1787 and continued to develop the orchard as part of the landscape plan (Figure 5.17). The property was inherited by three more generations of Adamses, who continued to experiment, develop, and maintain the orchard as part of the house grounds. The Adams orchard is a contributing feature within a larger historic property.

Figure 5.16: *(facing page)* Period plan of the Moses Cone Estate along the Blue Ridgeway Parkway, NC, showing a landscape design that integrated orchards with a system of carriage roads, pasture, forests, lakes and mountain views (Ian Firth, from Moses H. Cone Memorial Park Cultural Landscape Report, 1993).

Figure 5.17: *(above)* Existing conditions site plan of the Adams property at Adams National Historical Park showing the orchard as one component of a designed landscape, MA (K. Lacy, from Adams National Historic Site Cultural Landscape Report, 1997).

Table 5.7: Table identifying National Register criterion D (source: National Register Bulletin 15: How to Apply the National Register Criteria for Evaluation).

Applying National Register Criterion D to Orchards and Fruit Trees

Criterion	Type of Significance
D.	Having yielded or may be likely to yield, information important in prehistory or history

Criterion D:

A group of fruit trees or a single fruit tree may be significant because they contain archeological evidence that helps us understand the history or prehistory of America. The national parks have numerous examples of sites of former occupation, where the only tangible evidence of human intervention is the presence of fruit trees, either dead or alive. The fruit trees may be engulfed by forest vegetation, be standing or laying dead within a clearing or a field, or appear as stumps cut long ago. The field or cleared area may contain tree wells or depressions where former trees grew. The field, clearing, or forest may contain archeological evidence of former occupation, but the fruit trees themselves may also have the potential to yield valuable information.

Living or dead, fruit trees can yield information about the period in which the trees were planted or sown, and their tree form can provide information about species, the use of the site, the knowledge or skill of the users, and their life ways. Living trees can be identified by variety, (primarily through the morphology of fruits, flowers, or tree form) and this has the potential to yield information about the heritage or ethnicity of the growers, their origins, and their intentions in the use of the fruit, among others. Cut stumps or dead trees may have their species identified by their bark or wood, and dendrochronology can reveal the age of the trees. Archeological investigation based on a research design can yield information from pollen or phytolith analysis, and remote sensing techniques, such as soil conductivity, resistivity, or magnetometry may reveal the former locations of fruit trees in the soil.

Criterion D Example

At Hampton National Historic Site, a maintained meadow known to have contained an orchard in the late 18th and 19th centuries was recently found to have sprouting clumps of apple shoots after the cessation of frequent mowing activities. Ten to 15 large clumps of sprouts appeared to be regularly spaced, and are being investigated as the potential sprouting rootstocks of historic fruit trees. The meadow has the potential to yield important information about the former orchard. The rootstocks are currently being allowed to grow and bear fruit in order to aid in further identification.

In summary, orchards or fruit trees in the national park system may have more than one type of significance and may reflect more than one period. Orchards or fruit trees may be individually significant and may be listed in the National Register as a historic district or site. Many orchards or fruit trees are not individually significant as historic districts or sites, but are contributing features to the significance and integrity of a larger cultural landscape listed as a district or a site. As a feature of a larger landscape, orchards and fruit trees can provide a great deal of information about how the landscape was historically used, and the tastes, practices, and habits of the occupants. Their preservation allows for the most complete depiction of the landscape during the significant period, and can indicate the particular horticultural system that was viable at the time.

Defining the Period of Significance

The period of significance is the period in which the orchard, group of fruit trees, or single fruit tree attained historic significance, in accordance with National Register guidelines. This is the period from which the resource dates, or the period that the resource accurately represents. This distinction is made because orchards and fruit trees are living organisms with finite lifespans. At the initial point where an orchard, a group of fruit trees, or a single fruit tree is identified as a cultural resource, the majority of trees should date from the period of significance. Some trees in the group, however, may have been replaced since the period of significance but still accurately represent the significant period by their type, form, pruning style, and layout. After identification and determination of National Register eligibility, the trees in a historic orchard or group may be gradually replaced-in-kind over time according to the ***Secretary of the Interior's Standards for the Treatment of Historic Properties***, yet still retain the significance and integrity of the resource (NPS, 1995).

For example, the Buckner Homestead Historic District of Lake Chelan National Recreation Area was listed on the National Register in 1988. The district contains approximately 12 acres of orchard, but also roads, pasture, a farmhouse, numerous outbuildings, and an irrigation system of hand-dug ditches. The property takes its significance largely from the orchard, which is an outstanding example of the third (or modern) period in the history of American fruit tree culture, and the orchard characterizes the property as an early 20th century homestead of the Stehekin Valley in the

North Cascades mountains. The period of significance is from 1875 to 1949. It includes the initial construction of the irrigation system and planting of the commercial orchard, the peak of commercial productivity during the 1930s and '40s, and the orchard's declining productivity following changes in the personal circumstances of the Buckner family. When the Buckner Homestead was determined eligible for National Register in 1988, approximately 260 of the 400 extant trees dated from the period of significance. Since then, the park has adopted an Orchard Management Plan with preservation guidelines and has carefully replaced-in-kind a number of the original trees, following the *Secretary of the Interior's Standards* without loss of significance and integrity of the orchard.

An orchard may have more than one period of significance if it represents more than one period or has more than one association. For example, the orchard at the Adams National Historic Site has existed in its present location since the 1780s when John Adams acquired the property. The orchard was actively cultivated and modified by the Adams family through four successive generations until 1927. The extant orchard belongs to four periods of significance, one for each period of stewardship by President John Adams, his son President John Quincy Adams, his grandson Charles Francis Adams, and his great grandson Brook Adams. The orchard most accurately depicts the last two periods of significance in the form and layout of trees. However, the orchard can also be associated with the earlier two periods through the varieties present that were favored by John Adams and John Quincy Adams, and through the authenticity of the location of the orchard and its relationship to the rest of the property.

Defining Integrity

The historic integrity of an orchard, group of fruit trees, or a single tree is a measure of physical authenticity, conveyed by extant characteristics or features that were present during the period of significance. The National Register has defined the concept of integrity as multifaceted by containing seven aspects. Collectively, the seven aspects provide the measure of authenticity through location, design, setting, materials, workmanship, feeling, and association (Table 5.8).

Landscape Characteristics

The seven aspects of integrity are conveyed in cultural landscapes through their extant landscape characteristics and constituent features. In historic sites without landscape characteristics, or in orchards with fruit trees that are contributing features, the aspects of integrity are conveyed through the extant features. Landscape characteristics are the broad tangible

The Seven Aspects of Integrity Applied to Orchards and Fruit Trees

Quality	Description
Location	This is the place where the orchard or fruit trees were sown or planted, and their distribution upon the land.
Design	This is the combination of elements that create the form, plan, space, structure and style of an orchard or fruit trees in a horticultural system.
Setting	This is the physical environment of the orchard or fruit trees, including the land forms, rivers or streams, naturally-occurring vegetation, climate, elevation and aspect.
Materials	These are the physical elements that were combined or deposited in a particular pattern or configuration to form the orchard or fruit trees, including the seedling or grafted plant materials, ground cover plant materials, stakes, fences, windbreak and ditch materials.
Workmanship	This is the physical evidence of the crafts of a particular culture of people during the period of significance, such as cultivation and care of an orchard (propagation, planting, pruning, fertilizing, irrigating and harvesting) and protection of an orchard (pest control, animal husbandry, staking, fencing, and windbreaks).
Feeling	This is the orchard or fruit trees' expression of the aesthetic or historic sense of the period of significance, evoked by sounds, smells, and the seasonal rhythm of horticultural activities, productivity and change.
Association	This is the direct link or clear relationship between the important historic event, person or distinctive characteristics of a period, and an orchard or fruit trees.

Table 5.8: Table showing the seven qualities of integrity, as defined by the National Register of Historic Places. Integrity is a measure of the physical authenticity of historic places, and can be applied to orchards and fruit trees (sources: National Register Bulletin 15: How to Apply the National Register Criteria for Evaluation and Ian Firth: "Biotic Cultural Resources in the National Park System").

Landscape Characteristics Applied to Orchards

Landscape Characteristic	Description
Natural Systems and Features	These are the natural aspects that influenced the development and resultant form of the orchard, such as climate, geology, geomorphology, hydrology and physiology.
Spatial Organization	This is the arrangement of elements creating the ground, vertical and overhead planes that define and create spaces in the orchard.
Land Use	This is the organization, form, and shape of the orchard in response to land use.
Cultural Traditions	These are the practices that influenced land use, patterns of division, building forms and the use of materials in the orchard.
Circulation	This is the spaces, systems and materials that constitute the systems for movement in the orchard.
Topography	This is the three-dimensional configuration of the orchard ground surface related to land use, and characterized by features and orientation.
Vegetation	This is the fruit trees, ground covers, windbreaks, pasture vegetation, and other woody and herbaceous plant materials, both indigenous and introduced.
Buildings and Structures	These are the three-dimensional constructs of the orchard, such as farmhouses, fruit storage barns, fruit cellars, pickers' cabins, packing sheds, and garages.
Cluster Arrangement	This is the pattern of nodes of clustered features in the orchard, such as buildings and structures, and rows or blocks of fruit species or varieties.
Small Scale Features	This is the small elements that provide detail and diversity combined with function and aesthetics, such as a windmill, fruit barrels or boxes, tree ladders, tree stakes, fences, and equipment or machinery for planting, mowing, tilling, pruning, spraying, fertilizing, fruit harvesting, packing or fruit storage.
Constructed Water Features	These are the built features and elements that utilize water for aesthetic or utilitarian functions in the orchard, such as a diversion dam, diversion channel, irrigation ditches, head gates, check dams, irrigation pipes, sprinklers, water storage tanks, ponds, reservoirs, berms and water pumps.
Views and Vistas	These are the features that create or allow for a range of vision in the orchard, which can be natural or designed and controlled.
Archeological Sites	These are the sites in the orchard containing surface and subsurface remnants related to historic or prehistoric use.

patterns or intangible processes that influenced the development of a cultural landscape, or were formed through its development. The NPS has defined a list of 13 possible landscape characteristics that can be found in any cultural landscape, including orchards (Table 5.9). Not all 13 landscape characteristics are found in every cultural landscape, and the characteristics are not mutually exclusive.

Table 5.9: *(facing page)* Table indicating the range of landscape characteristics that may be found in any cultural landscape, as defined by the National Park Service, adapted to orchard landscapes. Historic orchards and other cultural landscapes convey their integrity through their extant landscape characteristics (source National Park Service, Cultural Resource Management Guideline, No. 5, 1997).

By definition, a cultural landscape is composed of landscape characteristics and features. The landscape characteristics are inextricably related as a dynamic system or matrix that is manifest as the substance of the landscape. For example, in a vernacular cultural landscape, such as an agricultural or mining landscape where the characteristic land use is extant, land use is typically interrelated to the characteristic, spatial organization—the natural systems and features—and the circulation system. In turn, these characteristics can be found to be interrelated to the characteristic vegetation, topography, and type of buildings and structures, etc. The patterns and processes that are the landscape characteristics exist in a matrix of interrelatedness.

A cultural landscape that, by definition, retains integrity, will possess an extant system of landscape characteristics. Orchards may or may not meet this definition. Some orchards possess a system of landscape characteristics and others do not. A simpler orchard, group of fruit trees, or a single fruit tree are composed of features rather than landscape characteristics. Orchards that have landscape characteristics can be identified as cultural landscapes and may be listed in the National Register as historic districts or historic sites. Orchards without landscape characteristics are not cultural landscapes but can be listed in the National Register as historic sites or as the contributing features of broader cultural landscapes that are listed as districts or sites.

The distinction between orchards with landscape characteristics and those without is dependent upon complexity. Orchards with landscape characteristics were developed historically as more complex systems, and they retain this array of interrelationships. Orchards without landscape characteristics are simpler, singular entities, rather like isolated structures, and they lack interrelated systems such as circulation, constructed water features, buildings and structures, and small scale features. The latter differentiation also applies to groups of fruit trees or a single tree. These were historically sown or planted as a simple entity, without interrelated systems. Where groups of fruit trees or single trees have both individual significance and integrity, they can be listed on the National Register as historic sites. Where they lack individual significance but retain integrity, they can be listed as contributing features of a broader cultural landscape.

The integrity of a historic property such as an orchard, group of fruit trees, or a single fruit tree, is reflected by as many as seven aspects of integrity: location, design, setting, materials, workmanship, feeling, and association. In more complex properties, such as a more complex orchard, the seven aspects of integrity are evidenced through the remaining landscape characteristics. Not all of the seven aspects of integrity are always relevant in every historic property. The property type and the type of significance of the historic context influence the relevance of the various aspects of integrity. A property associated with an event, historical pattern, or person might retain all seven aspects of integrity. Integrity of design and workmanship, however, might not be as important to the significance, and might not be relevant if the property is a site. A property significant under Criterion C must retain those physical features or landscape characteristics that characterize the type, period, or method of construction the property represents. Retention of design, workmanship, and materials will be more important than location, setting, feeling, and association for orchards or groups of fruit trees eligible under Criterion C. Location and setting will be important however, for those orchards or groups of fruit trees whose design is a reflection of their immediate environment.

The distinction separating cultural landscapes that are historic districts from those defined as historic sites is dependent on size, complexity, and the number of buildings. The National Register classifies small landscapes without buildings or with a single building as historic sites, and large landscapes with a number of buildings as historic districts (National Register Bulletin 30, 2). The Buckner Homestead is listed in the National Register as a historic district. It is a cultural landscape composed largely of an orchard. As illustrated by the following contemporary site plan, the historic district is a complex matrix of interrelated systems or landscape characteristics. The orchard's landscape characteristics include spatial organization, land use, natural systems and features, circulation, vegetation, constructed water systems, buildings and structures, and small scale features.

Example of Orchard Landscape Characteristics:

Buckner Orchard, Lake Chelan National Recreation Area, North Cascades National Park Service Complex, WA (Figure 5.18).

Natural Systems and Features

In 1911 the Buckner family took advantage of land that had been cleared of forest by the former settler, William Buzzard, to develop their orchard. Buzzard had planned to homestead the land, and responded to the opportunity presented by this relatively flat area on the Stehekin Valley bottom, beside an oxbow of the Stehekin River, to clear the forest and create farmable land. The Stehekin Valley is a glacially carved valley in the North Cascades mountains, and the valley is contained by steep mountain slopes that soar thousands of feet high. Very little flat land exists in the area. In locating their orchard here, the Buckners responded to the opportunities presented by the natural systems and features of the area. These included the river terrace, the presence of the Stehekin River and Rainbow Creek for irrigation water, and the fine climate for apple growing presented by the high elevation, aspect and dry summers. The Buckner's response to these natural systems and features is still evident today (Figure 5.19).

Figure 5.18: Existing conditions site plan of the Buckner Orchard in North Cascades National Park Service Complex showing irrigation ditches lining the rows of fruit trees, WA (C. Oshkello and J. Breitling, 1999).

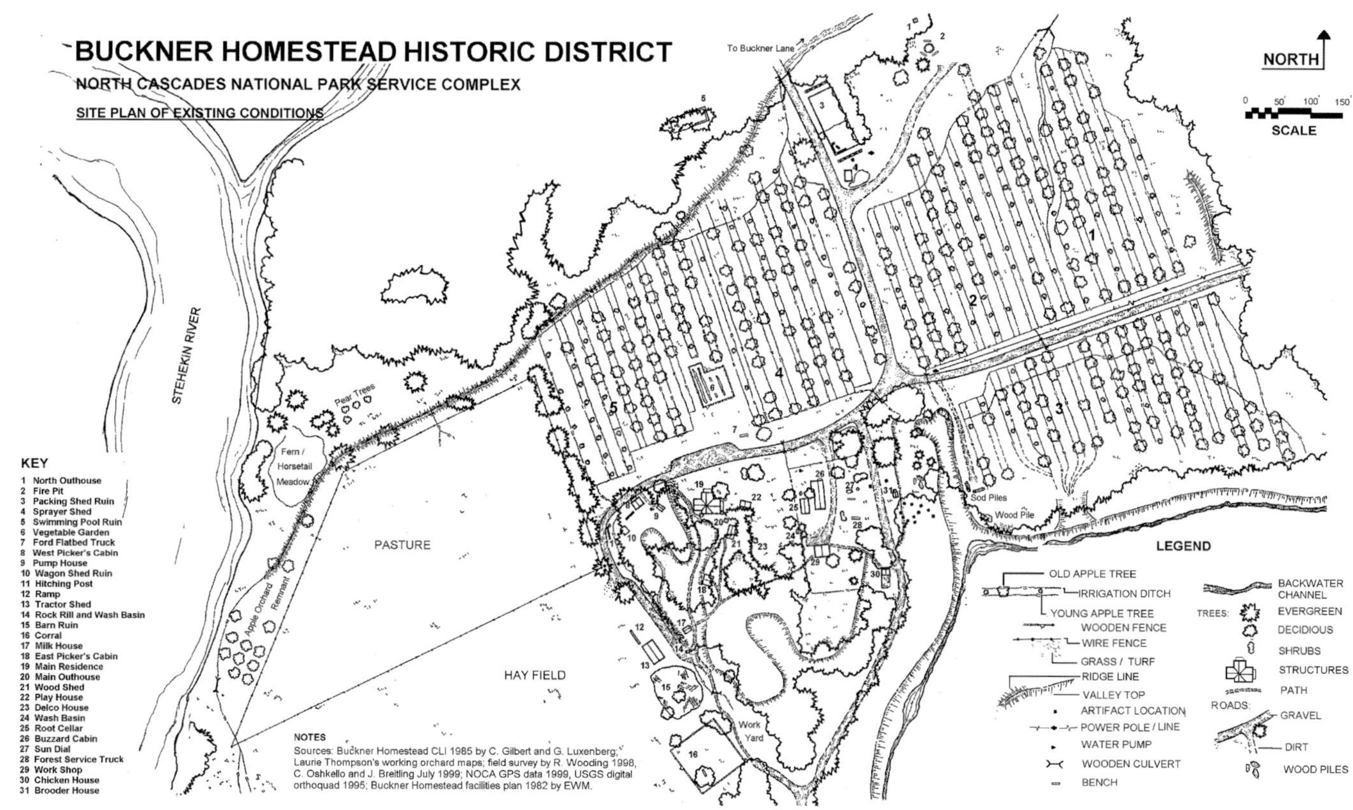

Figure 5.19: Aerial photograph of the Buckner orchard in the North Cascades National Park Service Complex showing the orchard's isolation on a terrace within an oxbow bend of the Stehekin River and surrounded by forest, WA (United States Geological Survey, 1990).

Spatial Organization

The Buckner orchard is laid out with an approximate cardinal orientation, generally north to south, and east to west, with essentially four quadrants separated by circulation. Three quadrants have orchards and the fourth quadrant has a building cluster, featuring the homestead residence and other outbuildings. The 12 acres of orchard trees are laid out on a grid of 30 x 30 feet spacing, typical for orchards planted from 1880 to 1945 (Figure 5.20). Rows and columns are aligned with a north-to-south, and east-to-west orientation. The dominant spaces are the overall forest clearing in which the orchard is located, the front yard area of the homestead residence, the hay pasture beside the orchard, and the intersection of two roads that bisect the orchard. Secondary spaces are found between the rows and columns of fruit trees, with the low canopy of the low-headed trees providing a ceiling plane overhead. This organization and pattern of spaces has existed since the period of significance.

Land Use

The dominant uses of the property were commercial apple production and as the Buckner family homestead. The orchard was the Buckner's main livelihood. Other uses of the property supported the Buckners or their orchard, including a residence with outbuildings and vegetable and flower garden area, a pasture for grazing livestock and horses (which provided manure for the orchard), a fruit storage area, and a water collection and

Figure 5.20: Oblique aerial photograph of the Buckner Orchard in North Cascades National Park Service Complex looking west, showing approximately five acres of commercial apple orchard dating to 1917, the historic circulation system, the horse pasture (in the distance), and the Stehekin River (far distance), WA (United States Geological Survey, 1990).

distribution system for irrigating the orchard. Today many of these uses are perpetuated by the park. The homestead is used for park housing and the orchard is tended by a park orchardist. Pack animals are grazed in the pasture, and the irrigation system is maintained to keep the orchard watered. The Stehekin community is welcomed by the park to harvest the apples and to attend an annual cider press celebration. Loss of the commercial use of the orchard has contributed to the demise of the packing shed and the absence of fruit boxes, and the absence of the Buckner family is evident in the loss of their traditions and life ways. But overall the continuing use of the property for apple orchard cultivation contributes considerably to its integrity.

Circulation

The Buckners built a number of two-track farm roads within their property, including two within the orchard. The roads were used for hauling in goods and supplies, and for hauling out packed fruit boxes for shipment to Chelan and Wenatchee. The roads also facilitated the movement of machinery and equipment within the orchard, for spreading manure, spraying nicotine (for pest control), mowing, harvesting apples, and adding new trees. Today, the alignment, scale, and character of the Buckner's roads remain unchanged and continue to serve the needs of the homestead and orchard.

Vegetation

The dominant palette of vegetation planted by the Buckners was apple trees, alfalfa, grasses for hay, and some shade trees and flowering plants around the homestead for ornament. This palette of vegetation remains today, and is highly indicative of the historic uses of the property. It remains in stark contrast to the vegetation surrounding the property: a native coniferous forest of cedar, fir, and pine. The apple trees are of the three dominant varieties planted by the Buckners: Common Delicious, Jonathan, and Rome Beauty, and these represent a typical combination of varieties for an eastern Washington orchard in the early 1900s. The trees were grafted onto Antonovka seedling rootstocks (allowing the trees to become full-sized), and were low-headed with trunks just 18-36 inches high.

The trees were pruned in the open-bowl style, with three main scaffold branches emanating from the top of the trunk. The orchard groundcover was primarily alfalfa, a typical "green manure" of the early 1900s commercial orchard. The non-native grasses sown in the pasture would provide forage for the Buckner's animals, and the ornamental plants and trees around the house decorated the domestic space. Today these patterns of vegetation remain extant, including the apple varieties and rootstock of the orchard trees, and alfalfa as the groundcover (though encroachment of sod grass has occurred). The pruning style of the apple trees is perpetuated by the park orchardist, and new replacement apple trees are custom-propagated to match the old trees.

Buildings and Structures

The property contains the Buckner's residence, the original William Buzzard log cabin, and a number of outbuildings that supported the Buckner's livelihood or their lifestyle in the Stehekin Valley, including two sleeping cabins for employee pickers, an electrical power generating house, a milk-separating house, a root cellar, a hay barn, a chicken house, a woodshed, a wagon shed, a packing shed (only the foundation is extant), and a playhouse and swimming pool for the Buckner's daughters. These buildings and structures remain today, and their materials, form, craftsmanship and use continue to characterize and typify the historic, vernacular uses of the property during the significant period (Figure 5.21).

Constructed Water Features

The existing conditions site plan shows the extent of the vast historic irrigation system within the orchard. Each column of apple trees has an irrigation ditch on both sides, watering the root zone of each apple tree from the east and west. Each irrigation ditch is fed by a feeder channel that is fed by a main channel, which is connected to the source: a diversion dam on Rainbow Creek. The flow of water is controlled by a system of head gates that redirect water to another part of the system, and these are opened or shut according to volume of flow and need for irrigation. The Buckners hand-dug this system to sustain their orchard and provide fresh water for

Figure 5.21: Photograph of the Buckner Homestead Cabin in the North Cascades National Park Service Complex built by pioneer William Buzzard in the early 1900s, then occupied and expanded by the Buckner family, who operated a commercial orchard at the homestead for approximately 50 years, WA (S. Dolan 1999).

Figure 5.22: *(above, left and right)* Photographs of the historic irrigation system at the Buckner orchard in North Cascades National Park Service Complex, fed by a main ditch from a diversion dam on Rainbow Creek to supply a system of minor ditches or furrows (left) that line each row of apple trees. The system is controlled by concrete head gates (right) that divert water to specific ditches as needed, WA (S. Dolan, 1999).

Figure 5.23: **2(right)** Historic photograph of a "furrow" irrigation system like the one in the Buckner orchard, modeled in a 1914 horticulture text book (from Lowther, 1914, courtesy of the Encyclopedia of Horticulture Cooperation).

the homestead (potable water was derived from a well). The main channel is a third of a mile long and contains some elevated wood channel sections to maintain a constant downhill gradient for gravity flow. The Buckners capitalized on the presence of the main channel as an aesthetic feature and created a trail (named Buckner Lane) with footbridges along it for evening strolls. Today the remarkable irrigation system is still intact and functioning, and many of the original features are extant (Figures 5.22 and 5.23).

Small Scale Features

The Buckner orchard contains an array of small scale features that date to the significant period, including a wagon, a tractor, a stump puller, a sprayer, a horse drawn mower, a horse-drawn harrow, a laundry tub, a stone fireplace, stone troughs, and stone garden ornaments. The features add detail and therefore great dimension to the authenticity of the property, and help enrich the early 20th-century feeling of the orchard.

Taken together these landscape characteristics comprise the horticultural system of the Buckner orchard. They convey the significance and integrity of the property, and qualified the property as eligible for listing in the National Register as a historic district in 1988.

Evaluating the Integrity of Individually Significant Versus Contributing Orchards and Fruit Trees

Most frequently, orchards, groups of fruit trees, or single fruit trees are listed in the National Register as features that contribute to the significance and integrity of larger historic districts or historic sites. In this case, the simpler orchard, group of fruit trees, or single fruit tree may lack the complexity of landscape characteristics, but may instead be composed of constituent features that convey the period of significance and are therefore very important to preserve. Orchards, groups of fruit trees, or single fruit trees that are included within historic districts or historic sites as contributing features typically lack individual distinction and may retain less integrity as a whole than an orchard or group of fruits trees that is individually eligible for listing in the National Register. Orchards, groups of fruit trees, or single fruit trees that are contributing features have an important role in conveying the significance and integrity of a historic property and should be treated as valuable cultural resources according to NPS CRM policy and the *Secretary of the Interior's Standards*.

Integrity of Rare Examples

Where orchards or groups of fruit trees have lost some integrity through physical alterations or loss of features over time, comparison with other similar properties may be needed in order to evaluate integrity. Comparison is important in deciding what physical features are essential to orchards or groups of fruit trees of that type. Comparative information is particularly important to consider when evaluating the integrity of an orchard, group of fruit trees or a single tree that is a rare surviving example of its type. The property must have the essential physical features that enable it to convey its historic context or historic information. The rarity and poor condition, however, of other examples of the type may justify accepting a greater degree of alteration or fewer features, provided that enough of the property survives for it to be considered a significant resource.

Examples of National Register Nominations

The Buckner Homestead Historic District National Register nomination, National Register Information System (NRIS) number 88003441, is a useful reference example of a nomination of an orchard as a historic district. The district's areas of significance are Agriculture and Exploration/Settlement. Another example of a nomination of an orchard, or rather a system of orchards as a historic district, is the Fruita Rural Historic District, NRIS 97000246. The district's areas of significance are Agriculture, Architecture, and Exploration/Settlement. The district is located within the Fremont Valley of south, central Utah, in Capitol Reef National Park.

A model example of a National Register nomination of an orchard property as a historic district is the Dorris Ranch Historic District nomination, NRIS 88000724. The district's area of significance is Agriculture. The 109-acre historic district near Springfield, Oregon is individually significant as the first commercial hazelnut (filbert) orchard in the United States and was instrumental in the establishment of a filbert industry in the Pacific Northwest. The district is composed of 75 acres of filbert orchards, planted in blocks of various spacing and varieties between 1903 and 1936. The 1988 nomination provides an excellent analysis and evaluation of the significance and integrity of the orchards in association with a well prepared historic context.

Examples of National Register nominations of properties with orchards as contributing features include the Bandelier CCC Historic District in New

Mexico, NRIS 87001452. The areas of significance for this rustic park village are Architecture and Social History. Another nomination example is the Adams National Historic Site in Massachusetts, NRIS 66000051, where the areas of significance for this residential landscape of the presidential Adams family are Landscape Architecture, Architecture, Literature, and Politics/Government. The Keys Desert Queen Ranch in California, NRIS 75000174, is another example of a nomination for a property where an orchard is a contributing feature. The area of significance of this desert homestead and ranch are Agriculture and Industry. The significance of each of these properties is not derived from their orchard or fruit trees, but instead, these resources make a very important contribution to the overall historic context and significance of the property.

Defining Boundaries

Definition of boundaries is an important step in the identification of an orchard or group of fruit trees as a historic district, or the identification of a group of fruit trees or a single fruit tree as a historic site. The definition of district or site boundaries should encompass the fullest extent of extant landscape characteristics or features that existed during the period of significance, or the fullest extent of the property with historic information potential (in the case of significance under Criterion D). An important characteristic or feature of an orchard, group of fruit trees, or a single tree is the area of land that was cultivated (in extant orchards, this is referred to as the orchard floor). Generally, boundaries should include the fullest extent of the area formerly cultivated with fruit trees, if the area is still recognizable as formerly cultivated land. For example, boundaries may encompass land where some fruit trees are now missing, but the land still exists as a cleared area and was historically part of the cultivated area of the orchard, group of fruit trees, or a single tree. The boundaries may include formerly cultivated land that is becoming reforested or populated with encroaching vegetation, if fruit trees still remain in the encroached-upon area and help to define the former extent of the orchard or cultivated area.

The boundaries should incorporate the boundaries of the orchard or group of fruit trees during the period of significance, to the extent that the historic boundaries are relevant to the significance and still retain integrity. The boundaries may be defined around contiguous but separately owned land parcels, through mutual consent or partnership agreement, to incorporate the fullest extent of the surviving historic property. However, in cases with a lack of consent, it may be necessary to define the boundaries just around the parcels with consensual ownership.

Introducing Cultural Resource Management of Orchards and Fruit Trees

Orchards, groups of fruit trees and single fruit trees that are eligible or potentially eligible for listing in the National Register of Historic Places are classified by the NPS as cultural resources. NPS cultural resource management (CRM) policy is defined by the *National Park Service Cultural Resource Management Guideline* (NPS CRM 97, 9), which outlines a process for management involving research, analysis and evaluation, leading to identification, stabilization, treatment, and subsequent preservation maintenance of cultural resources. Stabilization, treatment and preservation maintenance philosophies are based on the *Secretary of the Interiors Standards for the Treatment of Historic Properties and Guidelines for the Treatment of Cultural Landscapes* (NPS, 1992).

Identification

The process of identifying historic orchards and fruit trees involves historical research, field work, and documentation, followed by analysis and evaluation of significance and integrity. The products of initial identification may be an inventory document, such as the NPS Cultural Landscape Inventory (CLI), or a Determination of Eligibility form (DOE). The CLI and DOE are both mechanisms for obtaining consensus determinations of National Register eligibility through the State Historic Preservation Office (SHPO). A SHPO consensus determination may lead to the preparation of a National Register nomination and subsequent listing through the Keeper in the National Register of Historic Places.

Preservation Maintenance

Beyond initial identification, NPS CRM policy calls for the protection and preservation of cultural resources using the techniques of preservation maintenance, repair and replacement-kind, to perpetuate the same design, scale, form, and materials over time. Preservation guidelines are derived from the *Guidelines for the Treatment of Cultural Landscapes*. Preservation maintenance objectives for a specific orchard or fruit trees may be detailed in an Orchard Management Plan. This type of plan describes the history, significance, and existing conditions of an orchard or fruit trees, and can define management objectives and describe maintenance regimens for preservation, or to implement restoration or rehabilitation treatments. Preservation maintenance actions for a historic orchard or fruit trees may involve all or some of the following activities: winter and/or summer pruning, weeding, aerating, mowing, cultivating, mulching, integrated pest management, fruit thinning, fruit harvesting, irrigating, fertilizing, monitoring, and documenting activities. Preservation maintenance activities should be performed by qualified personnel with training in the cultural resource values of the orchard or fruit trees and the management objectives.

Stabilization

When a historic orchard, a group of fruit trees, or a single fruit tree is in poor or unstable condition, NPS CRM policy calls for interim stabilization actions to be taken until the resource can receive preservation maintenance or ultimate treatment. Stabilization is an interim step involving actions of temporary longevity to prevent the further deterioration of the condition of the resource. Stabilization actions in historic orchards or fruit trees may involve deadwood removal, bracing, sucker removal, encroaching vegetation removal, brush-hogging, mowing or aerating the orchard floor, irrigating and mulching. Germplasm conservation may also be performed as part of stabilization. Evaluation by a qualified orchard specialist should be performed before stabilization actions are undertaken.

Treatment

When the cultural resource management objectives of an orchard or fruit trees involve treatments other than preservation maintenance, a treatment plan is prepared. Four types of treatment are defined by NPS CRM policy and the *Secretary of the Interior's Standards:* preservation, restoration, rehabilitation, and reconstruction. A treatment plan defines the type of treatment and provides recommendations or actions needed to implement treatment. The treatment plan for an orchard or fruit trees may be part of a Cultural Landscape Report (CLR), which describes the history, significance, and integrity of a cultural landscape (Part I) and provides a treatment plan for the landscape (Part II). The treatment plan for an orchard or fruit trees that are part of a larger cultural landscape should be consistent and compatible with the treatment, significance, and period of significance of the landscape as a whole. An Orchard Management Plan may be prepared after a CLR to provide more detailed information on the implementation of a treatment plan.

The *Guidelines for the Treatment of Cultural Landscapes* provide standards for the four types of treatment.

- Preservation standards require retention of the greatest amount of historic fabric, including the landscape's historic form, features, and details as they have evolved over time.
- Restoration standards allow for the depiction of a landscape at a particular time in its history by preserving materials from the period of significance and removing materials from other periods.

- Rehabilitation standards acknowledge the need to alter or add to a cultural landscape to accommodate continuing or new uses while retaining the landscape's historic character.
- Reconstruction standards establish a framework for recreating a vanished or non-surviving landscape with new materials, primarily for interpretive purposes.

The treatment plan for an orchard or fruit trees should be based on clearly defined management objectives that are compatible with the type and level of significance of the property. The plan may include drawings and construction details for the removal and replacement of fruit trees or encroaching vegetation, or for the re-establishment of orchard floor vegetation, circulation routes, boundary demarcations or other features. Treatment plans are prepared and implemented by personnel who meet the qualifications required by the *Secretary of the Interior's Standards.* Restoration and rehabilitation treatment actions for orchards and fruit trees may involve the propagation of replacement trees using cuttings of appropriate scion wood and potential grafting with appropriate rootstock material. Restoration or rehabilitation treatments may involve re-planting with custom-propagated replacement trees and the follow-up care of browse and sunscald protection, and irrigation until establishment.

Germplasm Conservation

After the implementation of treatment, NPS CRM policy calls for the ongoing protection and preservation of the orchard or fruit trees as cultural resources, through the resumption of cyclic preservation maintenance and monitoring activities, to be documented over time. Germplasm conservation is recommended as a long-term preservation strategy for a significant orchard, group of fruit trees or single fruit tree.

Germplasm conservation preserves the genes of each variety and each species (full complement of genotypes) present, in perpetuity. Conservation can be achieved by two means: one, through a living collection of trees representing all of the genotypes of the orchard or fruit trees and maintained off-site, such as in a plant nursery, and two, through cryogenic means, involving the USDA National Plant Germplasm Repositories. Cryogenically conserved germplasm is plant tissue held at sub-zero temperatures in liquid nitrogen, which can be thawed later and used to propagate replacement trees.

Germplasm conservation uses fruit tree cuttings from the scion. All scions of the same variety have the same genotype and therefore it is unnecessary to conserve the germplasm of every scion in an orchard or a group of fruit trees. Instead, germplasm conservation should focus on preserving *each variety within each species present.* Germplasm cuttings are taken from dormant shoots with several replicates (multiple individuals) of the same species and same variety during the dormant season. Conservation using the USDA National Plant Germplasm Repositories will involve the development of a partnership agreement with the agency.

PART III:
FURTHER INFORMATION

National Parks with Fruit Trees More than 50 Years Old

(Adapted from *Inventory and Conservation of Genetic Resources in the Form of Historically Significant Fruit and Nut Trees in the National Park System,* by William Coli and Nora Mitchell, 1992.)

Northeast Region

Acadia National Park
Adams National Historical Park
Booker T. Washington National Monument
Boston National Historical Park
Cape Cod National Seashore
Delaware Water Gap National Recreation Area
Edison National Historic Site
Eisenhower National Historic Site
Eleanor Roosevelt National Historic Site
Fire Island National Seashore
Fredericksburg and Spotsylvania National Military Park
Friendship Hill National Historic Site
George Washington Birthplace National Monument
Gettysburg National Military Park
Hampton National Historic Site
Home of Franklin D. Roosevelt National Historic Site
Hopewell Furnace National Historic Site
Martin Van Buren National Historic Site
Minuteman National Historical Park
Morristown National Historical Park
New River Gorge National River
Richmond National Battlefield Park
Sagamore Hill National Historic Site
Salem Maritime National Historic Site
Saratoga National Historical Park
Saugas Iron Works National Historic Site
Shenandoah National Park
Springfield Armory National Historic Site
Statue of Liberty National Monument
Theodore Roosevelt Inaugural National Historic Site
Valley Forge National Historical Park
Vanderbilt Mansion National Historic Site
Women's Rights National Historical Park

National Capital Region

Antietam National Battlefield
Catoctin Mountain Park
Chesapeake and Ohio Canal National Historical Park
Colonial National Historical Park
Manassas National Battlefield Park
Oxon Hill Farm
Piscataway Park
Prince William Forest Park

Southeast Region

Andersonville National Historic Site
Andrew Johnson National Historic Site
Big South Fork National River
Biscayne National Park
Blue Ridge Parkway
Canaveral National Seashore
Carl Sandburg Home National Historic Site
Chickamauga and Chattanooga National Military Park
Cowpens National Battlefield Park
Cumberland Gap National Historical Park
Cumberland Island National Seashore
Fort Donelson National Battlefield
Fort Frederica National Monument
Fort Matanzas National Monument
Fort Pulaski National Monument
Great Smoky Mountains National Park
Jean Lafitte National Historical Park and Preserve
Kennesaw Mountain National Battlefield Park
Mammoth Cave National Park
Natchez Trace Parkway
Ninety Six National Historic Site
Ocmulgee National Monument
Shilo National Military Park

Midwest Region

Apostle Islands National Lakeshore
Buffalo National River
Cuyahoga Valley National Park
Effigy Mounds National Monument
Fort Scott National Historic Site
George Washington Carver National Monument
Herbert Hoover National Historic Site

Homestead National Monument of America
Hot Springs National Park
Indiana Dunes National Lakeshore
Isle Royale National Park
Lincoln Boyhood National Memorial
Lincoln Home National Historic Site
Ozark National Scenic Riverways
Pea Ridge National Military Park
Pictured Rocks National Lakeshore
Pipestone National Monument
Saint Croix National Scenic Riverway
Sleeping Bear Dunes National Lakeshore
William Howard Taft National Historic Site
Wilson's Creek National Battlefield
Wind Cave National Park

Intermountain Region

Aztec Ruins National Monument
Bandelier National Monument
Big Bend National Park
Bighorn Canyon National Recreation Area
Canyon de Chelly National Monument
Capitol Reef National Park
Carlsbad Caverns National Park
Chiricahua National Monument
Colorado National Monument
Coronado National Monument
Dinosaur National Monument
Fort Davis National Historic Site
Glacier National Park
Glen Canyon National Recreation Area
Grand Canyon National Park
Guadalupe Mountains National Park
Hubbell Trading Post National Historic Site
Lyndon B. Johnson National Historical Park
Organ Pipe Cactus National Monument
Pecos National Historical Park
Pipe Spring National Monument
Tumacacori National Monument
Salinas Pueblo Missions National Monument
Zion National Park

Pacific West Region

Channel Islands National Park
Death Valley National Park
Eugene O' Neil National Historic Site
Fort Vancouver National Historic Site
Golden Gate National Recreation Area
Great Basin National Park
Hawaii Volcanoes National Park
John Day Fossil Beds National Monument
John Muir National Historic Site
Joshua Tree National Monument
Kalaupapa National Historical Park
Lake Roosevelt National Recreation Area
Lewis and Clark National Historical Park
Manzanar National Historic Site
Mojave National Preserve
Nez Perce National Historical Park
North Cascades National Park Service Complex
Olympic National Park
Point Reyes National Seashore
Pu'uhonua o Honaunau National Historical Park
Redwoods National Park
San Juan Island National Historical Park
Santa Monica Mountains National Recreation Area
Sequoia and Kings Canyon National Park
Whiskeytown National Recreation Area
Whitman Mission National Historic Site
Yosemite National Park

List of Repositories

Smithsonian Institution
Horticulture Branch Library.
Arts and Industries Building
9 Jefferson Drive, SW
Washington, DC, 20560

Adams Library
Adams National Historical Park
135 Adams Street
Quincy, Massachusetts 02169-1749
Curator 617 773-1177
www.nps.gov/adam

Massachusetts Horticultural Society Library
Elm Bank Horticulture Center
900 Washington Street (Route 16)
Wellesley, Massachusetts 02482-5725
Thomas Herrera-Mishler, Executive Director
617 933-4955
www.masshort.org

National Archives at College Park
8601 Adelphi Road
College Park, Maryland 20740-6001
www.archives.gov/dc-metro/college-park

United States Department of Agriculture (USDA)
National Agricultural Library
10301 Baltimore Avenue
Beltsville, Maryland 20705-2351
301 504-5755
www.nal.usda.gov

USDA
National Germplasm Resources Laboratory
10300 Baltimore Boulevard
Room 103, Building 003, BARC-West
Beltsville, Maryland 20705
Gary Kinard, Acting Research Leader
301 504-6235

USDA, Agricultural Research Service
National Plant Germplasm System
http://sun.ars-grin.gov/npgs

National Plant Germplasm Repository for Pear – Corvallis, Oregon
USDA, Agricultural Research Service
33447 SE Peoria Road
Corvallis, Oregon 97333-2521
Kim Hummer, Research Leader, Curator
541 738-4200
cor@ars-grin.gov

National Plant Germplasm Repository for Citrus and Dates – Riverside, California
USDA, Agricultural Research Service
1060 Martin Luther King Boulevard
Riverside, California 92507-5437
Robert Krueger, Curator
951 827-4399
rivrk@ars-grin.gov

National Plant Germplasm Repository for Tree Fruit/Nut Crops and Grapes – Davis, California
USDA, Agricultural Research Service
One Shields Avenue, UCD
Davis, California 95616-8607
Ed Stover, Curator
530 752-7009
dav@ars-grin.gov

National Plant Germplasm Repository for Subtropical Horticulture – Citrus
Subtropical Horticultural Research Station
USDA, Agricultural Research Service
13601 Old Cutler Road
Miami, Florida 33158
Tomas Ayala-Silva, Raymond Schnell, Curators
305 254-3635
mia@ars.grin.gov

National Plant Germplasm Repository for Tropical Fruit and Nuts
USDA, Agricultural Research Service
P.O. Box 4487
928 Stainback Highway
Hilo, Hawaii 96720
Francis Zee, Curator
808 959-5833
fzee@pbarc.ars.usda.gov

National Collection of Genetic Resources for Pecans and Hickories
USDA, Agricultural Research Service
10200 FM 50
Somerville, TX 77879
L.J. Grauke, Curator
979 272-1402
brwlg@ars-grin.gov

National Plant Germplasm Repository for Apples
USDA, Agricultural Research Service
Plant Genetic Resources Unit
Cornell University Experiment Station
630 W. North Street
Geneva, New York 14456-0462
Philip Forsline, Acting Research Leader
315 787-2390
Philip.Forsline@ars.usda.gov

New York Horticultural Society Records
The LuEsther T. Mertz Library
The New York Botanical Garden
200th Street and Kazimiroff Boulevard
Bronx, New York 10458-5126
718 817-8604
http://library.nybg.org

Pennsylvania Horticultural Society McLean Library
100 N. 20th Street
Philadelphia, Pennsylvania, 19103-1495
Reference Librarian: 215 988-8782
www.pennsylvaniahorticulturalsociety.org
Relevant Organizations

Relevant Organizations

American Horticultural Society
River Farm
7931 East Boulevard Drive
Alexandria, Virginia 22308
703 768-5700
800 777-7931
www.ahs.org

American Pomological Society
University of Pennsylvania
103 Tyson Building
University Park, Pennsylvania 16802
aps@psu.edu
http://americanpomological.org

American Society for Horticultural Science
www.ashs.org

Brogdale Horticultural Trust
National Fruit Collection
Brogdale Farm Road
Faversham
Kent, ME13 8X2
United Kingdom
From U.S. 011 44 1795-535286
www.brogdale.org

California Rare Fruit Growers
www.crfg.org

East of England Apples and Orchards Project
www.applesandorchards.org.uk

Foundation Plant Services
University of California – Davis
1 Shields Avenue
Davis, California 95616-8600
530 752-3590
fpms@ucdavis.edu
www.fpms.ucdavis.edu

Fruit and Nut Inventory
Seed Savers Exchange
Rural Route Three
P.O. Box 239
Decorah, Iowa 52101
319 382-5990

Home Orchard Society
www.homeorchardsociety.org

International Society for Horticultural Science
www.ishs.org

Mount Vernon Research and Extension Unit
Washington State University
16650 State Route 536
Mount Vernon, WA 98273
http://mtvernon.wsu.edu

Indiana Nut Growers Association
www.nutgrowers.org

International Fruit Tree Association
P.O. Box 5006
Wenatchee, WA 98807
www.ifruittree.org

Midwest Fruit Explorers
www.midfex.org

National Park Service
Olmsted Center for Landscape Preservation
Boston National Historical Park
Charlestown Navy Yard
Quarters C
Boston, Massachusetts 02129
www.nps.gov/oclp

Nebraska Nut Growers Association
University of Nebraska
122 Mussehl Hall
Lincoln, Nebraska 68583-0716

North American Fruit Explorers
Jill Vorbeck
Route One, P.O. Box 94
Chapin, Illinois 62628
217 245-7589
www.nafex.org

North Nut Growers Association
www.northernnutgrowers.org

NRSP5/IR-2 Virus-Test Fruit Tree Collection
Irrigated Agriculture and Research Center
Washington State University
24106 North Bunn Road
Prosser, Washington 99350
(509) 786 9251
bhowell@tricity.wsu.edu
www.nrsp5.wsu.edu

Thomas Jefferson Center for Historic Plants
Monticello
P.O. Box 316
Charlottesville, Virginia 22902
(804) 984 9821
www.monticello.org

Western Cascade Fruit Society
http://www.geocities.com/wcfsfruit

Western Washington Fruit Research Foundation
www.wwfrf.org

Worcester County Horticultural Society
Tower Hill Botanical Garden
11 French Drive
P.O. Box 598
Boylston, MA 01505-0598
www.towerhillbg.org

Glossary

Cultivar

The abbreviated term for a "cultivated variety." A man-made variation within a species. The name of the cultivar follows the genus and species and is denoted by single quotation marks. The initial letters of the cultivar name are capitalized, e.g., *Pyrus communis* 'Winter Bartlett,' and the cultivar name is not italicized (see also: "variety").

Clonal dwarfing rootstock

A commercial horticulture term for dwarfing rootstocks that are clones or genetically identical to each other. Dwarfing rootstocks are cloned to perpetuate desirable characteristics and to guarantee a rootstock's ability to confer these characteristics upon the scion, such as extent of dwarfness, disease resistance, and youthful bearing of fruit.

Clone

The scion portion of a tree propagated by grafting. A clone is genetically identical to the parent. Clones, as opposed to seedlings, do not have genetic variation (see also: "scion").

Cultural resource

Term for a building, site, district, object, or structure evaluated as historically significant.

Dwarfing rootstock

A rootstock that limits the height of a grafted tree to be shorter than the standard height (see also: "dwarf tree").

Dwarf tree

A tree grown on a rootstock that limits its final height to be shorter than the standard height. Dwarf trees are generally classified as semi-standard, about two thirds of standard height; semi dwarf, about half of standard height; and dwarf, about one third of standard height.

Espalier

A method of training and pruning trees in which branches are trained horizontally in a single plane, usually against a wall, fence or trellis.

Evaluation

National Register term for the process by which the significance and integrity of a historic property are judged and eligibility for listing in the National Register of Historic Places is determined.

Germplasm

The genetic material, especially its specific molecular and chemical constitution that forms the physical basis of heredity and is transmitted from one generation to the next. When applied to plants, it is the term given to seed or any vegetative material from which plants can be propagated.

Grafting

A method of propagation in which two different plants are joined together in order to take advantage of the special characteristics of each (see also: "rootstock" and "scion").

Graft union

The joint between the two parts of the grafted tree which have grown together. When visible, the union appears as a line, scar, indent, or change in bark pattern on the tree trunk. The height of the graft union on the trunk has varied over time, and during the 19th century, was commonly buried at planting (see also: "grafting").

Historic context

National Register term for an organizing structure for interpreting history that groups information about historic properties which share a common theme, common geographical area, and a common time period. The development of historic contexts is a foundation for decisions about the planning, identification, evaluation, registration, and treatment of historic properties, based upon comparative historic significance.

Historic integrity

National Register term for the unimpaired ability of a property to convey its historical significance. Integrity is a measure of the physical authenticity of a historic property or cultural resource.

Historic significance

National Register term for the value or importance of a historic property within the patterns of American history, in relation to a historic context. Significance may be in association with important events or persons, or for importance in design or construction, or for information potential.

Interstem

A grafting method in which the rootstock and scion are joined by an intermediate graft, known as the interstem (see also: "grafting").

Low headed tree

The term for a tree with a scaffold borne upon a short trunk. The head or point of attachment of the main branches to the trunk is set by pruning in the first or second year after planting. The practice of low heading or creating fruit trees with a low head on a short trunk, was used to control height in the transition from standard to dwarf trees between 1881 and 1945.

National Register of Historic Places

The Nation's official list of cultural resources worthy of preservation. Authorized under the National Historic Preservation Act of 1966, the National Register is part of a national program to coordinate and support public and private efforts to identify, evaluate, and protect our historic and archeological resources. Properties listed in the Register include districts, sites, buildings, structures, and objects that are significant in American history, architecture, archeology, engineering, and culture. The National Register is administered by the National Park Service, which is part of the U.S. Department of the Interior.

Naturalize

A term used to describe the ability of some plant species to sexually reproduce outside of their native ecosystems, i.e., their ability to sexually reproduce themselves as non-native or exotic plants.

Perry

An alcoholic beverage made of fermented pear juice. It is similar to cider, in that it is made using a similar process and often has a similar alcoholic content, around 8.5% alcohol by volume. The word is derived from the French, poiré.

Pomology

The scientific study and cultivation of fruits, particular tree fruits.

Registration

National Register term for the process by which a historic property is documented and nominated or determined eligible for listing in the National Register of Historic Places.

Rootstock

The term used in grafting to refer to the root system (see also: "scion").

Scaffold

The framework of major branches growing from the trunk on a tree.

Scion

The term used in grafting to refer to the upper portion of the graft, typically the aerial portion of the grafted tree.

Seedling

A tree grown from seed.

Sport

A genetic variation on a part of a tree, such as a limb, with different characteristics, such as redder fruit.

Spur-type variety

A strain of a variety manifested as trees with more fruit-bearing spurs and fewer vegetative shoots than the parent variety.

Standard tree

A tree grown on its own roots or grafted to a seedling rootstock that allows the tree to reach its natural height (see also: "dwarf tree").

Strain

A variation within a specific variety or a specific cultivar (see also: "variety" and "cultivar").

Variety

A naturally occurring variation within a species. The variety name is a Latin name written after the genus and species. The variety name is italicized along with the genus and species, e.g., *Prunus cerasifera atropurpurea* (see also: "cultivar").

Vegetative propagation

The process of producing a new plant from a portion of another plant, such as a stem or a branch. Also known as asexual reproduction, the process does not involve the mixing of genes from different parents as in sexual reproduction. The new offspring is genetically identical or a clone of the parent.

Whip

The term for an unbranched young tree typically one to two years old.

Bibliography

American Pomological Society. *Catalogue of Fruits Recommended for Cultivation in the Various Sections of the United States and the British Provinces by the American Pomological Society*. Washington, D.C.: Government Printing Office, 1897. Revised 1899, 1909.

Appleland News (Monthly). Nos. 1-17, Jan. 10, 1936-Dec. 1952. Yakima, Washington: FW Shields.

Apple Research Digest. Nos. 1-153; May 1946-Sept. 1959. Yakima, Washington: Washington State Dept. of Agriculture in cooperation with the Washington State Apple Commission and the Institute of Agricultural Sciences, Washington State College.

Auchter, Eugene Curtis. *Orchard and Small Fruit Culture*. New York: J. Wiley & Sons, Inc., London: Chapman & Hall, Ltd., 1932.

Bailey, L.H. *Sketch of the Evolution of our Native Fruits*. New York: The Macmillan Company; London: Macmillan & Co., Ltd., 1898.

———. *The Apple Tree*. New York: The Macmillan Company; London: Macmillan & Co. Ltd., 1922.

———. *The Principles of Fruit Growing, with Applications to Practice*. New York: The Macmillan Company; London: Macmillan & Co. Ltd., 1915.

Barritt, Bruce H. Intensive Orchard Management: *A Practical Guide to the Planning, Establishment and Management of High Density Apple Orchards*. Yakima, Washington: Good Fruit Grower, 1992.

Barritt, Bruce H. Interview by Susan Dolan, February 7, 2001. Horticulturist, Washington State University Tree Fruit Research and Extension Center, Wenatchee, Washington.

Beach, Spencer Ambrose. *The Apples of New York*. 2 v. New York (State) Agricultural Experiment Station, Geneva. Report, 1902/03, [pt.] II. Albany, New York: J.B. Lyon Company, printer, 1905.

Benmayor, Rina and Andor Skotnes, eds. *Migration and Identity*. Oxford; New York: Oxford University Press, 1994.

Bercaw, Louise Oldham. *The Apple Industry in the United States: a Selected List of References on the Economic Aspects of the Industry Together with some References on Varieties*. Washington, D.C.: United States Dept. of Agriculture, Bureau of Agricultural Economics, 1927.

Birnbaum, Charles, and Chris Capella Peters. *The Secretary of the Interior's Standards for the Treatment of Historic Properties with Guidelines for the Treatment of Cultural Landscapes*. Washington, D.C.: United States. Dept. of the Interior, National Park Service, 1997.

Blagrave, Joseph. *The Epitome of the Art of Husbandry: comprising all necessary directions for the improvement of it, viz.: plowing, sowing, grafting, gardening, ordering of flowers, herbs, directions for the use of the angle, ordering of bees, together with the gentleman's heroick exercise, discoursing of horses, their nature and use, with their diseases and remedies, of oxen, cows, calves, sheep, hogs, with the manner of ordering them, their diseases and remedies : of the nature of marle, the best way of planting clover-grass, hops, saffron, liquorish, hemp, etc. : to which is annexed by way of appendix a new method of planting fruit trees and improving of an orchard, with several other new editions.* 2nd ed. London: Printed for Benjamin Billingsley, 1670.

Brackett, G.B. *The Pear and How to Grow It.* Washington, D.C.: United States Dept. of Agriculture, farmers' bulletin no. 482, 1912.

Bradley, Richard. *The Fruit Garden Display'd, setting forth the several varieties of fruit ripe in every month of the year...* London: Printed for W. Mears and J. Stone, 1732.

Bregger, John Taylor. *Conserving Soil and Moisture in Orchards and Vineyards.* Washington, D.C.: United States Dept. of Agriculture, bulletin, 1945.

Bright, A.C. *Apples Galore! The History of the Apple Industry in the Wenatchee Valley.* 1st ed. Wenatchee, Washington: Directed Media, Inc., c1988.

Brookes, Reid Merrifield. *Register of New Fruit and Nut Varieties, 1920-1950.* Berkeley, California: University of California Press, 1952.

Browning, Frank. *Apples.* New York, North Point Press, 1998.

Budd, Joseph L. *American Horticultural Manual.* New York: J. Wiley & Sons, 1903-11.

Bultitude, John. *Apples: A Guide to the Identification of International Varieties.* Seattle, Washington: University of Washington Press, 1983.

Burford, Thomas. *Apples: A Catalog of International Varieties.* Monroe, Virginia, 24575, revised 2007.

Bush, Carroll D. *Nut Grower's Handbook, A Practical Guide to the Successful Propagation, Planting, Cultivation, Harvesting and Marketing of Nuts.* New York: Orange Judd Company, 1941.

Calhoun, Creighton Lee. *Old Southern Apples.* Blacksburg, Virginia: McDonald & Woodward Publishing Company, 1995.

California State Commission of Horticulture. *Apple Growing in California.* California: State Printing Office, 1914.

Carlsen, Earl W. *Apple Handling Methods and Equipment in Pacific Northwest Packing and Storage Houses.* Washington, D.C.: United States Dept. of Agriculture, Production and Marketing Administration, Marketing and Facilities Research Branch, 1953.

Carlson, Robert F., ed. *North American Apples: Varieties, Rootstocks, Outlook*. East Lansing, Michigan: Michigan State University, 1970.

Cobbett, William. *An Epitome of Mr. Forsyth's Treatise on the Culture and Management of Fruit Trees: also, notes on American gardening and fruit, with designs for promoting the ripening of fruits and securing them as family comforts, and further, of economical principles in building farmers' habitations by an American farmer.* Philadelphia, Pennsylvania: Wm. Poyntell & Co., 1804.

Coit, John Eliot. *Citrus Fruits.* New York: The Macmillan Company, 1915.

Cole, S.W. *The American Fruit Book; containing directions for raising, propagating, and managing fruit trees, shrubs and plants; with a description of the best varieties of fruit, including new and valuable kinds..* Boston, Massachusetts: J.P. Jewett; New York: C.M. Saxon, 1849.

Coli, William M., and Nora Mitchell. "Inventory and Conservation of Genetic Resources in the Form of Historically Significant Fruit and Nut Trees in the National Park System". Prepared under a Cooperative Agreement between the University of Massachusetts, Amherst, Extension Service and the National Park Service, North Atlantic Regional Office, Cultural Landscapes Program, December 1992.

Cook, Moses. *The manner of Raising, Ordering, and Improving Forest and Fruit Trees: also, how to plant, make and keep woods, walks, avenues, lawns, hedges, &c., with several figures in copper-plates, proper for the same: also rules and tables shewing how the ingenious planter may measure superficial figures, with rules how to divide woods or land, and how to measure timber and other solid bodies, either by arithmetick or geometry shewing the use of that most excellent line, the line of numbers, by several new examples, with many other rules, useful for most men / by Moses Cook ..; whereunto is now added that ingenious treatise of Mr. Gabriel Plattes, viz. A discovery of subterranean treasure.* London: Printed for Peter Parker, 1679.

Cooper, Jacob Calvin. *Walnut Growing in Oregon.* Portland Oregon: Passenger Department, Oregon Railroad and Navigation Company, Southern Pacific Lines in Oregon, 1910.

Copeland, Robert Morris. *Country Life: A Handbook of Agriculture, Horticulture and Landscape Gardening.* 5th edition, revised. Boston, Massachusetts: Dinsmoor and Company, 1866.

Coxe, William. *A View of the Cultivation of Fruit Trees, and the Management of Orchards and Cider; with accurate descriptions of the most estimable varieties of native and foreign apples, pears, peaches, plums, and cherries, cultivated in the middle states of America: illustrated by cuts of two hundred kinds of fruits of the natural size.* Philadelphia, Pennsylvania: M. Carey and Son, Nov 1, 1817.

Daniel Smith and Company. *A Catalogue of Fruit Trees for Sale, in the fall of 1806, by Daniel Smith and Company, at Burlington, state of New Jersey, near Philadelphia.* [Microfilm]. Philadelphia, Pennsylvania: Printed at the office of the United States Gazette, [1806].

Dolan, Susan A. "Orchard Management Plan for John Muir National Historic Site." United States Dept. of Interior, National Park Service, Pacific West Region, July 2006.

Downing, Andrew Jackson. *Cottage Residences; or, A Series of Designs for Rural Cottages and Cottage Villas, and their Gardens and Grounds.* 4th edition. New York: J. Wiley, 1852.

Downing, Andrew Jackson. *The Fruits and Fruit Trees of America: or, The culture, preparation, and management, in the garden and orchard, of fruit trees generally; with descriptions of all the finest varieties of fruit, native and foreign, cultivated in this country.* 7th edition. New York: Wiley and Putnam, 1847.

Downing, Andrew Jackson. *A Treatise on the Theory and Practice of Landscape Gardening, Adapted to North America; with a View to the Improvement of Country Residences, with Remarks on Rural Architecture.* New York: Orange Judd company, c. 1859.

Emmonds, Ebenezer. *Agriculture of New York.* Albany, New York: C. Van Benthuysen & Co., 1846-54.

Ferree, David C. *A History of Fruit Varieties: the American Pomological Society: One Hundred and Fifty Years, 1848-1998.* Yakima, Washington: Good Fruit Grower Magazine, 1998.

Firth, Ian. "Biotic Cultural Resources in the National Park System". Washington, D.C.: United States Dept. of Interior, National Park Service, 1983.

Firth, Ian. *Moses H. Cone Memorial Park Cultural Landscape Report.* United States Dept. of Interior, National Park Service, Southeast Regional Office, Cultural Resources Planning Division, August 1993.

Fishman, Ram. *The Handbook for Fruit Explorers.* Chapin, Illinois: North American Fruit Explorers, 1986.

Folger, J.C., and S.M. Thomson. *The Commercial Apple Industry of North America.* New York: The Macmillan Company, 1923.

Fortier, Samuel. *Irrigation of Orchards.* Washington, D.C.: United States Dept. of Agriculture, Farmers' Bulletin No. 404, 1910.

Fuller, Andrew Samuel. *The Nut Culturist: a treatise on the propagation, planting and cultivation of nut-bearing trees and shrubs, adapted to the climate of the United States.* New York: Orange Judd Company, 1896.

Gardner, Victor Ray. *The Cherry and its Culture.* New York: Orange Judd Company, London: K. Paul, French, Grubner & Co. Ltd., 1930.

———, Frederick Charles Bradford and Henry Daggett Hooker, Jr. *The Fundamentals of Fruit Production.* 1st edition. New York: McGraw-Hill Book Company, 1922.

Gellatly, John Arthur. *A History of Wenatchee; the Apple Capital of the World.* Wenatchee, Washington: 1963.

Gould, H.P. *Apple-Orchard Renovation.* Washington, D.C.: United States Dept. of Agriculture, bulletin, 1922.

Gourley, J.H. *Textbook of Pomology.* New York: The Macmillan Company, 1922.

Greenoak, Francesca. *Forgotten Fruit: the English Orchard and Fruit Garden.* London: A. Deutsch, 1983.

Haley, W.D. "Johnny Appleseed; a Pioneer Hero." *Harpers New Monthly Magazine* 830-836. New York: November 1871.

Harwood, W.S. *New Creations in Plant Life: An Authoritative Account of the Life and Work of Luther Burbank.* London: The Macmillan Company, 1914.

Hatch, Peter J. *The Fruits and Fruit Trees of Monticello.* Charlottesville, Virginia: University Press of Virginia, 1998.

Hedrick, U.P., assisted by G.H. Howe. *The Pears of New York.* Report of the State of New York Department of Agriculture, Agricultural Experiment Station for the year 1921, 29th annual report, vol. 2, pt. II. Albany, New York: J. B. Lyon Company, 1921.

———. *A History of Agriculture in the State of New York.* New York: Hill and Wang, 1966.

———. *A History of Horticulture in America to 1860.* New York: Oxford University Press, 1950.

———. *A History of Horticulture in America to 1860; with an addendum of books published from 1861 to 1920 by Elisabeth Woodburn.* Portland, Oregon: Timber Press, 1988.

———. *Cyclopedia of Hardy Fruits.* New York: The Macmillan Company, 1922.

———. *Systematic Pomology.* New York: The Macmillan Company, 1925.

Hovey, C. M. *The Fruits of America, containing richly colored figures, and full descriptions of all the choicest varieties cultivated in the United States... Forty-eight richly colored plates.* New York: D. Appleton & Company, 1853.

Hume, H. Harold. *Citrus Fruits.* New York: The Macmillan Company, 1957.
———. *The Pecan and Its Culture.* Petersburg, VA: The American Fruit and Nut Journal, 1906.

International Society for Horticultural Science. "Performance of the NC-140 Cherry Rootstock in North America." ISHS Horticulture 451: III International Cherry Symposium. September, 2005. www.actahort.org
International Society for Horticultural Science. "Recent Developments in Cherry Rootstock Research". ISHS Horticulture 468: III International Cherry Symposium. September, 2005. www.actahort.org

Kains, Maurice Grenville. *Food Gardens for Defense.* New York, Greenberg, 1942.

Kennedy, C. Todd. "Survey of the Orchards of Yosemite National Park." Report prepared for Yosemite National Park, Resource Management, 1991.

Kenrick, William. *The New American Orchardist; or an account of the most valuable varieties of fruit, of all climates, adapted to cultivation in the United States, with their history, modes of culture, management, uses, and the culture of silk. With an appendix on vegetables, ornamental trees, shrubs and flowers.* Boston, Massachusetts: Russell, Odiorne, and Metcalf, 1835.

Klose, Nelson. *America's Crop Heritage: The History of Foreign Plant Introduction by the Federal Government.* Ames, Iowa: The Iowa State College Press, 1950.

Knapp, Halsey B. *Growing Tree and Small Fruits.* New York: J. Wiley & Sons, Inc., London, Chapman & Hall, Ltd., 1929.

Kopolow, Carol. *Fruit Orchards: January 1984-April 1991.* Beltsville, Maryland: National Agricultural Library, [c.1991].

Jacques, Georges. *Practical Treatise on the Management of Fruit Trees with Descriptive Lists of the Most Valuable Fruits for General Cultivation Adapted to the Interior of New England.* Worcester, Massachusetts: Erastus N. Tucker, 1849.

Jaynes, Richard A., Ed. *Nut Tree Culture in North America.* Hamden, Connecticut: Northern Nut Growers Association, 1979.

Lape, Fred. *Apples and Man.* New York: Van Nostrand Reinhold, 1979.

Larsen, R. Paul. *That We May Eat. "The quiet revolution in the apple orchard."* Washington, D.C.: United States Government Printing Office, United States Dept. of Agriculture Yearbook p. 158-168, 1975.

Lawson, William. *A New Orchard and Garden; or, the best way for planting, grafting, and to make any ground good for a rich orchard: particularly in the north, and generally for the whole common-wealth... with the country-housewifes garden for herbs of common use... as also, the husbandry of bees, with their several uses and annoyances: all being the experience of forty and eight years of labour, ereunto is newly added the art of propagating plants. The fifth time corrected and much enlarged.* London: Printed by W. Wilson, for E. Brewster and George Sawbridge, 1656.

Loudon, J.C. (John Claudius). *Encyclopedia of Agriculture.* 5th edition. London: Longman, Brown, Green, and Longmans, 1844.

Loudon, J.C. (John Claudius). *Encyclopedia of Gardening.* London: Longman and Company, 1826.

Loudon, J.C. (John Claudius), ed. *The Landscape Gardening and Landscape Architecture of the Late Humphrey Repton, Esq., being his entire works on these subjects.* London: printed for the editor, and sold by Longman and Company, 1840.

Lowther, Granville. *The Encyclopedia of Practical Horticulture; a reference system of commercial horticulture, covering the practical and scientific phases of horticulture, with special reference to fruits and vegetables.* North Yakima, Washington: The Encyclopedia of Horticulture Corporation, [c.1914], reprinted 1929.

Macomber, Alvin Zohe. *Summary of Trade and Tariff Information: Deciduous Tree Fruit and Fruit Products: TSUS (A) items 146.10-146.24, 146.90-147.02, 148.70-148.86, 149.18-149.28, 150.02-150.50, 152.05-152.42, 152.62, 152.78, 152.8840, 153.02-153.32, and 154.05-154.90.* Washington, D.C.: United States International Trade Commission, [1983].

Mascall, Leonard. *A booke of the art and maner, howe to plante and graffe all sortes of trees [microform]: how to set stones and sow pepynes to make wylde trees to graffe on, as also remedies and medicines : with diverse other newe practises.* Imprinted at London: by Henry Bynneman for John Wight, 1584.

Masumoto, David Mas. *Four Seasons in Five Senses, Things Worth Savoring.* New York: W.W. Norton and Company, Inc., 2003.

McClelland, Linda Flint, J. Timothy Keller, P. Genevieve Keller, and Robert Z. Melnick. *National Register Bulletin 30: Guidelines for Evaluating and Documenting Rural Historic Landscapes.* Washington, D.C.: United States Dept. of Interior, National Park Service, 1990.

Morgan, Joan and Alison Richards. *The Book of Apples.* London: Ebury Press (with Brogdale Horticultural Trust), 1993.

Phillips, Becky, contributing writer. "Innovation, Specialization Grow with World Apple Market." *WSU Today*, Friday October 29, 2004, Pullman, Washington: News Bureau, Washington State University, 2004. http://www.wsutoday.wsu.edu

Pollan, Michael. *The Botany of Desire: A Plant's Eye View of the World.* Toronto: Random House, Inc., 2001.

Postman, Joseph. "The Endicott Pear Tree. Oldest Living Fruit Tree in North America." National Plant Germplasm Repository, December 2002. http://www.ars-grin.gov/cor/pyrus/endicott.pear.html

Powell, Edward Payson. *The Orchard and Fruit Garden.* New York: Doubleday, Page & Co., 1910.

Price, Robert. *Johnny Appleseed: Man and Myth.* Indiana: Indiana University Press, 1954.

Prince, William Robert. The Pomological Manual. New York: T. & J. Swords; Philadelphia, J. Dobson, 1831.

Quinn, Patrick T. *Pear Culture for Profit.* New York: The Tribune Association, 1869, revised 1883.

Ragan, William Henry. *Nomenclature of the Pear; a catalogue-index of the known varieties referred to in American publications from 1804 to 1907.* Washington, D.C.: Government Printing Office, 1908.

Reisner, Mark. "Mark's Fruit Crops." University of Georgia, Department of Horticulture, September 2006. www.uga.edu/fruit.html

Robertson, Adele Crockett. *The Orchard: A Memoir.* New York: Bantam Books, 1997.

Rolfs, Peter Henry. *Propagation of Citrus Trees in the Gulf States.* Washington, D.C.: United States Dept. of Agriculture, Bulletin No. 539, 1913.

Rom, Roy C. and Robert F. Carlson, eds. *Rootstocks for Fruit Crops.* New York: Wiley, 1987.

Rosenstein, Mark. *In Praise of Apples: A Harvest of History, Horticulture and Recipes.* Asheville, North Carolina: Lark Books, 1996.

Rossi, William, ed. *"Wild Apples" and Other Natural History Essays by Henry David Thoreau.* Athens, Georgia: University of Georgia Press, 2002.

Royal Horticultural Society (Great Britain). *Fruit: Present and Future.* London: R.H.S., 1966.

———. *The Fruit Garden Displayed.* London: R.H.S., 1951.

Sanders, Rosanne. *The Apple Book.* New York: Philosophical Library, Inc., 1988.

Saxton, C.M. *The American Fruit Book.* New York: Saxton, Barker and Co., 1860.

Scofield, David and Angus Kress Gillespie. *Rooted in America: Foodlore of Popular Fruits and Vegetables.* 1st edition, Knoxville, Tennessee: University of Tennessee Press, c1999.

Slingerland, K., and B. Lay. "Plum Cultivars – European and Japanese." *Fact Sheet AG Index 216.* Ontario: Ministry of Agriculture and Food, August 2002.

Smith, Daniel and Company. *A Catalogue of Fruit Trees for Sale, in the Fall of 1806, by Daniel Smith and Company, at Burlington, State of New Jersey, near Philadelphia.* Philadelphia, Pennsylvania: Printed at the office of the United States Gazette [1806].

Smith, Muriel, W.G. *The National Apple Register of the United Kingdom.* London: Ministry of Agriculture, Fisheries and Food, 1971.

Stebbins, R. "Choosing Pear Rootstocks for the Pacific Northwest." *Pacific Northwest Extension Publication 341.* United States Department of Agriculture, January 1995.

Stewart, Henry. *Irrigation for the Farm, Garden and Orchard.* New York: Orange Judd Company, 1877.

Stoll, Steven. *The Fruits of Natural Advantage: Making the Industrial Countryside in California.* Berkeley, California: University of California Press, 1998.

Strong, William Chamberlain. *Fruit Culture and the Laying Out and Management of a Country Home.* Boston, Massachusetts: New York, Houghton, Mifflin and Co., 1885.

Taylor, Judith M., M.D. *The Olive in California. History of an Immigrant Tree.* Berkeley, California: Ten Speed Press, 2000.

Teskey, Benjamin J.E. and James S. Shoemaker. *Tree Fruit Production.* 3rd edition, Connecticut: The AVI Publishing Company, Inc., 1978.

The Northwest Farm & Orchard (Monthly). Vol. 1, no. 7. Spokane, Washington: Farm and Orchard Publishing Company, Feb. 1910.

Thomas, Jon Jacob. *The American Fruit Culturist.* New York: W. Wood and Company, 1897.

Tobin, Mark E. and Milo E. Richmond. *Vole Management in Fruit Orchards.* Washington, D.C.: United States Dept. of the Interior, Fish and Wildlife Service: Springfield, Virginia National Technical Information Service (NTIS) [distributor, 1993].

Todd, Sereno Edwards. *The Apple Culturist. A Complete Treatise for the Practical Pomologist.* New York: Harper & Brothers, 1871.

Tukey, Harold Bradford. *Dwarfed Fruit Trees for Orchard, Garden, and Home: with special reference to the control of tree size and fruiting in commercial fruit production.* Ithaca, New York: Cornell University Press, 1979, c1964.

———. *The Pear and Its Culture; design for both the amateur and the commercial grower.* New York: Orange Judd Company, London: K. Paul Trench, Trubner & Co. Ltd., 1928.

Twiss, Sally. *Apples: A Social History.* London: The National Trust, Centurion Press Ltd., 1999.

United States Department of Agriculture. *Establishing and Managing Young Apple Orchards.* Prepared by National Program Staff, Agricultural Research Service. Revised Dec. 1976. Washington, D.C.: United States Dept. of Agriculture, 1976.

United States Department of Interior, National Park Service. *Cultural Resource Management Guideline.* Washington, D.C.: United States Dept. of Interior, National Park Service, 1997.

———. *National Register Bulletin 15: How to Apply the National Register Criteria for Evaluation.* Washington, D.C.: United States Dept. of Interior, National Park Service, 1985.

———. *Secretary of the Interior's Standards for the Treatment of Historic Properties.* Washington, D.C.: United States Dept. of Interior, National Park Service, revised 1995.

———. *Secretary of the Interior's Standards for the Treatment of Historic Properties and Guidelines for the Treatment of Cultural Landscapes.* Washington, D.C.: United States Dept. of Interior, National Park Service, 1992.

United States Department of the Treasury. *Report to Congress on the Depreciation of Fruit and Nut Trees.* Washington, D.C.: Dept. of the Treasury, 1990.

United States Division of Pomology. *Nut Culture in the United States, Embracing Native and Introduced Species.* Washington, D.C.: Government Printing Office, 1896.

Venette, Nicolas. *Art de tailler les arbres fruitiers. The Art of Pruning Fruit-Trees: with an explanation of some words which gardiners make use of in speaking of trees, and a tract, Of the use of the fruits of trees for preserving us in health or for curing us when we are sick / translated from the French original set forth the last year by a physician of Rochelle.* London: Printed for Thom. Basset, 1685.

Washington Agricultural Statistics Service. *Washington Fruit Survey, 1986.* Prepared by Washington Agricultural Statistics Service. Olympia, Washington: Issued cooperatively by State of Washington, Dept. of Agriculture, United States Dept. of Agriculture, National Agricultural Statistics Service, 1987.

Waugh, Frank Albert. *Dwarf Fruit Trees.* New York: Orange Judd Company, 1906.

———. *Plums and Plum Culture; a monograph of the plums cultivated and indigenous in North America, with a complete account of their propagation, cultivation and utilization.* New York: Orange Judd Company, 1901.

———. *Systematic Pomology, Treating Description, Nomenclature, and Classification of Fruits.* New York: Orange Judd Company, 1903.

———. *The American Peach Orchard; a sketch of the practice of peach growing in North America at the beginning of the twentieth century.* New York: Orange Judd Company, 1913.

———. *The American Apple Orchard; a sketch of the practice of apple growing in North America at the beginning of the twentieth century.* New York: Orange Judd Company, 1908.

Way, Roger D. *Apple Cultivars Introduced by the New York State Agricultural Experiment Station, 1914-1968.* Geneva, New York: New York State Agricultural Experiment Station, 1971.

Webber, H.J. *Yearbook of the United States Department of Agriculture for 1896.* Washington, D.C.: United States Department of Agriculture, 1896.

Weiss, Harry Bischoff. *The History of Applejack or Apple Brandy in New Jersey from Colonial Times to the Present.* Trenton, New Jersey: Agricultural Society, 1954.

Wheeler, Bob. "Fruit Trees for Interior Alaskan Landscapes." *Forest Canopy* website. Fairbanks, Alaska: University of Alaska at Fairbanks Cooperative Extension, March 1999. http://www.uaf.edu/coopext/forestry/canopy/canopyjul99.htm

Wilder, Marshall P. *Descriptive Catalog of Fruit and Ornamental Trees, Grape Vines, Shrubs, Roses, Greenhouse and Herbaceous Plants, Cultivated and For Sale by Marshall P. Wilder.* Dorchester, Massachusetts: Wilder, Marshall P., 1858.

Williams, Max W. "Pear Production." *Agriculture Handbook, no. 526.* Washington, D.C.: United States Dept. of Agriculture, Science and Education Administration, 1978.

Wilson, David Scofield and Angus Kress Gillespie. *Rooted in America: Foodlore of Popular Fruits and Vegetables.* Knoxville, Tennessee: University of Tennessee Press, 1999.

Wilson, Robert Wayne. *A History of the Washington State Apple Advertising Commission.* Washington: Washington State Apple Commission, 1966.

Yepsen, Roger B. *Apples.* New York: W.W. Norton, 1994.

Index

A

B

C

L

M

N

O

P

U.S. GOVERNMENT PRINTING OFFICE: 2009 — 579-511 / R3328 Region No. 10